College Prep Math Workbook

Practice Exercises for College Algebra Success

M.J. Sanders

2015

Those who strive to give expression to the gift of living through virtue and learning will never regret the choices that they make.

ISBN-13: 978-1511433334

ISBN-10: 1511433337

Cover Photo: The photo on the cover depicts a section of the Talmadge Memorial Bridge. The bridge spans 1100 ft of the Savannah River between downtown Savannah, Georgia and the South Carolina Lowcountry. Completed in November 1991, the cable-stayed bridge is dedicated to Eugene Talmadge, who served as the democratic Governor of Georgia from 1933-37 and 1941-43. Its total length is 1.9 miles.

TABLE OF CONTENTS

Foreword: The exercise sets contained Part I are designed to provide short daily practice in mathematics skills such as factoring, solving equations, understanding and using function notation, working with exponents and radicals, etc. It is recommended that you complete only one exercise set each day in order to provide the continued practice over a period of time. The time spent on each exercise set varies and is dependent upon your skill and familiarity with the concepts presented, but should require about 15-30 minutes per day. The subject matter is meant to be cyclic with repeated emphasis on skills that are fundamental for college mathematics success. For best utilization, it is recommended that you attempt to complete the exercise sets over a 30-day period. However, short breaks for weekends or other responsibilities should not provide a severe impediment provided that you are persistent in filling in any knowledge gaps when they arise.

Daily work notes are provided in Part II that speak directly to the pertinent aspects of each day's exercise set. If you are unsure of how to complete exercises for a given day, please refer to these work notes for guidance. The directions, review notes, and hints provided, offer guidance where you might require a brief refresher. Brief and to the point, with examples when needed for clarity, these notes may help your understanding of the topics presented by the exercises if you are having difficulty. Depending upon your preparedness, you may find you need to consult another text or online resources for additional guidance if a topic seems foreign.

Part III contains a complete answer set. It is recommended that you work the daily exercise sets to the best of your ability and gain extra help by reviewing the work notes in Part II before reviewing the answer set. You should attempt to ascertain the difference between your responses and the answer set when they arise. For example, many exercises ask for the complete factorization of an expression over the real numbers. If your answer differs, it may be because more factorization is possible or it may be, simply, that the two answers are written in slightly different forms. You should be diligent about accounting for any discrepancies.

I wish you great success in your college mathematics career and hope that this work helps you succeed.

M.J. Sanders

PART I: Daily Exercise Sets

Day 1

1 Daily Drill on Fundamentals: Multiplying Algebraic Expressions

Expand the following by multiplying factors.

1. $(x)(2x + 3)$

    $2x^2 + 3x$

~~4.~~ $(x)(x - 3)(x + 1)$

    $1x^3 - 3x^3 + 1x$
    $x^3 - 2x^2 - 3x$

2. $(x^2)(x^2 + 4)$

    $x^4 + 4x^4$

~~5.~~ ? $(2x - 4)(3x + 1)$

    $6x^2 - 10x - 4$

~~3.~~ $(x - 2)(x + 5)$ $x^2 + 3x - 10$ ~~6.~~ $(x - 1)^2$ $x^2 - 2x + 1$

    $1x^2 2x + 5x$ $-1x^2$

2 Solving Algebraic Equations

1. Solve $2x - 5 = 0$

    $x = 5/2$

2. Solve $(x)(3x - 2) = 0$

    $x^2 = 0$

~~3.~~ Solve $(x^2)(x + 7)(4x - 3) = 0$

    $x = 0$ or $x = -7$ or $x = 3/4$

3   Factor Expansion

Expand the following by multiplying factors.

1. $(x-3)(x+3)$

$x^2 - 9$

4. $(x+4)^3$

$5x^3$

2. $(2x+5)(2x-5)$

$4x^2 - 25$

5. $(3x+2)^3$

$5x^3$

3. $(2x-3)^2$ $4x^2 - 12x + 9$

$-1x^2$

6. $(x+1)^4$

$2x^4$

4   Math To Go

Answer the following, then explain.

1. True or False: The solutions to $(x)(x-3) = 0$ are $x = 0$ or $x = 3$. Explain your reasoning.

2. True or False: The solutions to $(x)(x-3) = 5$ are $x = 5$ or $x = 2$. Explain your reasoning.

Day 2

1   Daily Drill on Fundamentals: Multiplying Algebraic Expressions

Expand the following by multiplying factors.

1. $(3x)(2 - 5x)$ 4. $(x + 3)(x - 3)(x + 4)$

2. $(x - 1)^2(x^2 + 1)$ 5. $(2x - 1)(3x - 2)(x)$

3. $(2x - 1)\left(\dfrac{x + 3}{2}\right)$ 6. $(3x - 1)^2$

2   Solving Algebraic Equations

1.   Solve $3x - 1 = 4$

2.   Solve $(x - 3)(3x + 5) = 0$

3.   Solve $(3x - 2)(x^2)\left(\dfrac{32 - 8x}{3}\right) = 0$

3   More Factor Expansion

Expand the following by multiplying factors.

1. $(x - 1)(x - 2)(x - 3)$ 4. $(x - 1)^4$

2. $(x + 5)^2(x - 5)$ 5. $(x + 1)^5$

3. $(3x + 3)^3$ 6. $(1 - x)^4$

4   Math To Go

Draw a figure corresponding to the following word problem and carefully label any appropriate quantities with variables and/or numbers. Write down an equation to solve the problem. Solve the problem being sure to state your solution using the appropriate units.

A rancher has 100 feet of fencing to create a rectangular corral.

1. If x denotes the length of the corral, write an expression in x that can be used to determine the width of the corral.

2. For what values of x is the expression you wrote valid? (That is, for what values of x does your formula make sense?)

Day 3

1　Daily Drill on Fundamentals: Factoring

Factor the following over the real numbers as much as possible.

1.　　$x^2 - x$

2.　　$x^2 - 9$

3.　　$x^3 + 2x^2$

4.　　$x^4 - 16$

5.　　$x^3 + 1$

6.　　$x^2 + 9x + 20$

2　Linear Equation Practice

1.　　Find the slope of the line passing through $(0, 0)$ and $(4, 12)$.

2.　　Find the equation of the line passing through $(0, 0)$ and $(4, 12)$.

3.　　Find the y-intercept of the line passing through $(0, 0)$ and $(4, 12)$.

3   Interval Notation

Write the following sets of real numbers using interval notation.

1. $\{x : x > 4\}$ 4. $\{x : 0 < x \leq 10\}$

2. $\{x : x \neq 5\}$ 5. $\{x : |x| > 4\}$

3. $\{x : x \leq 2 \text{ or } x > 5$ 6. $\{x : |x + 2| \leq 1\}$

4   Math To Go

Draw a figure corresponding to the following word problem and carefully label any appropriate quantities with variables and/or numbers. Write down an equation to solve the problem. Solve the problem being sure to state your solution using the appropriate units.

1. The Fédération Internationale de Football Association (FIFA) is an association governed by Swiss law founded in 1904. It sets regulations regarding football (referred to as soccer in the US). Field dimension requirements for international matches allow the length of a field to range between 100 m and 110 m and the width to range between 64 m and 75 m. How much larger is the largest possible field size (in sq. m) as compared to the smallest possible field size?

Day 4

1 Daily Drill on Fundamentals: Factoring

Factor the following over the real numbers as much as possible.

1. $x^6 + x^3$

4. $2x^2 - 9x - 5$

2. $x^2 - 5x + 6$

5. $x^3 - 1$

3. $x^4 + 3x^2$

6. $3x^3 - 5x^2 - 2x$

2 Linear Equation Practice

1.   Find the slope of the line passing through $(-2, 4)$ and $(1, 10)$.

2.   Find the equation of the line passing through $(-2, 4)$ and $(1, 10)$.

3.   Find the y-intercept of the line passing through $(-2, 4)$ and $(1, 10)$.

3   Interval Notation

Write the following intervals of real numbers using set-builder notation. For example,
$(-\infty, 1) = \{x : x < 1\}$.

1. $[5, 10]$ 4. $(-\infty, 0]$

2. $[-3, 4)$ 5. $(1, 2)$

3. $(-\infty, 0)$ 6. $[10, \infty)$

4   Math To Go

Draw a figure corresponding to the following word problem and carefully label any appropriate quantities with variables and/or numbers. Write down an equation to solve the problem. Solve the problem being sure to state your solution using the appropriate units.

1. A rectangular soccer field is twice as long as it is wide. If the perimeter of the field is 270 m, what are the dimensions of the field?

Day 5

1 Daily Drill on Fundamentals: Solving Linear Equations

Solve the following equations. (Recall that *solve* means find all values of all variables that satisfy the equation).

1. $2x + 3 = 11$

4. $3a - 8 = -14$

2. $3(x - 4) + 9 = 21$

5. $20t + 3(t - 2) = 5(6 - t) - 8$

3. $2x + 6(x - 10) = 3x$

6. $3x + 4y = 2(x + 1 + 2y)$

2 Linear Equations

1. Find the slope of the line given by the equation $y = 5x - 2$.

2. Find the slope of the line given by the equation $y - 3 = 12(x + 1)$.

3. Find the slope of the line given by the equation $4x - 3y + 9 = 0$.

3   Interval Notation

Write the following sets of real numbers using interval notation.

1. $\{x : x \le -5\}$ 4. $\{x : 0 > x \ge -4\}$

2. $\{x : x \ne 0\}$ 5. $\{x : |x| < 5\}$

3. $\{x : x \ge 2 \text{ or } x \le 0\}$ 6. $\{x : |x + 10| \ge 5\}$

4   Math To Go

Recall that the circumference, C, of a circle is given by the equation $C = 2\pi r$ where r denotes the radius of the circle. The circumference of a sphere is the circumference of an "equator" of the sphere.

1. The circumference of a FIFA-sanctioned soccer ball is between 68 and 70 cm. Find the radius of a soccer ball whose circumference is 68 cm (round your answer to the nearest tenth of a cm.)

2. Find a formula for the radius, r, of a circle whose circumference is C.

Day 6

1   Daily Drill on Fundamentals: Exponents and Notation

Find the value of the following expression if the quantity is defined in $\mathbb{R}$, the set of real numbers. If it is undefined, then say so.

1. $\sqrt{81}$ 4. $(-1)^{7/2}$

2. $81^{1/2}$ 5. $(-27)^{2/3}$

3. $8^{2/3}$ 6. $-9^{3/2}$

2   Rational Exponents

Write the following expressions using rational exponents.

1. $\sqrt{16^3}$

2. $(\sqrt{9})^3$

3. $\sqrt[n]{a^m}$

3   More rational exponents

Determine whether the following equalities are true or false.

1. $\sqrt{5} = 5^{1/2}$

4. $(\sqrt{7})^3 = \sqrt{7^3}$

2. $\sqrt[3]{5^2} = 5^{3/2}$

5. $49^{-1/2} = \dfrac{1}{7}$

3. $\sqrt[3]{5^2} = 5^{2/3}$

6. $\sqrt{x^2} = x$

4   Math To Go

Recall that the volume, V, of a sphere is given by the formula

$$V = \frac{4}{3}\pi r^3$$

where r denotes the radius of the sphere.

1. Find the radius of a soccer ball (to the nearest cm) given that it's volume is 33510 cubic cm.

2. Find a formula for the radius, r, of a sphere of given volume V.

Day 7

1 Daily Drill on Fundamentals: Solving Linear and Linear-like Equations

Solve the following equations if possible. (Recall that *solve* means find all values of all variables that satisfy the equation.)

1. $5x + 2 = 3$

4. $\dfrac{x + 2}{x + 1} = 4$

2. $3(2 - x) + 7 = 0$

5. $\dfrac{3}{x - 3} + \dfrac{2}{x + 1} = \dfrac{6x - 1}{x^2 - 2x - 3}$

3. $\dfrac{2}{x} = 3$

6. $\dfrac{2x}{x - 1} - \dfrac{1}{x - 1} = \dfrac{1}{x - 1}$

2 Function Notation Practice

Let $y = f(x) = \dfrac{12}{\sqrt{x-2}}$.

1. Find $f(6)$

2. Find $f(11)$

3. Find $f(t)$

3   Domain of a Function

Find the domains of the following functions and put your answers in interval notation.

1. $f(x) = \dfrac{5}{x-3}$

4. $f(x) = \dfrac{\sqrt{x-2}}{x-5}$

2. $f(x) = \sqrt{x-3}$

5. $f(x) = \sqrt[3]{x} - 2x$

3. $f(x) = \dfrac{3x-1}{x^2-x-6}$

6. $f(x) = \dfrac{3x+4}{2x^2+7}$

4   Math To Go

Draw a figure corresponding to the following word problem and carefully label any appropriate quantities with variables and/or numbers. Write down an equation to solve the problem. Be sure to state your solution using the appropriate units.

1. The height, h, (in m) of a soccer ball t seconds after being kicked by a goalie is given by $h(t) = -9.8t^2 + 30t + 1$. How high is the ball 2 seconds after being kicked?

Day 8

1   Daily Drill on Fundamentals: Equivalent expressions

Determine whether or not the given expressions are equivalent. Explain why or why not.

1. $\dfrac{x^2 - x - 6}{x - 3}$ and $x + 2$

2. $\dfrac{3x^2 - 27}{x - 3}$ and $\dfrac{3x^2 + 9x}{x}$

3. $\dfrac{(x - 4)(x^2 + 1)}{x^2 + 1}$ and $x - 4$

2   Function Notation Practice

Let $y = f(x) = x^2 - x + 3$.

1. Find $f(1)$

2. Find $f(t)$

3. Find and simplify $f(x + h)$

3   Domain of a Function

Find the domains of the following functions and put your answers in interval notation.

1. $f(x) = \dfrac{\sqrt{3}}{x^2 - 16}$

4. $f(x) = \dfrac{\sqrt[3]{x}}{2x - 5}$

2. $f(x) = \sqrt{2x - 1}$

5. $f(x) = \sqrt[4]{x^2 + 10}$

3. $f(x) = \dfrac{x^2 + x}{x^2 - 1}$

6. $f(x) = \dfrac{3x + 4}{\sqrt{2 - x}}$

4   Math To Go

In plain words, explain what the following equations mean in context. Use correct units to identify quantities.

1. Let $h(t)$ denote the height (in feet) of a sapling t years after being transplanted.

$$h(5) = 10$$

2. Let $C(t)$ denote the concentration of a certain drug in a patient's blood (measured in mg/dL) t minutes after injection.

$$C(10) = .5$$

M.J. Sanders

Day 9

1   Daily Drill on Fundamentals: Factoring

Factor the following over the real numbers as much as possible.

1. $3x^2 + 17x - 6$

4. $12x^3 - 96$

2. $-x^2 + x + 30$

5. $x^3 + x^2 - 3x - 3$

3. $x^4 - 5x^2 - 36$

6. $x^5 - x^3 - 6x$

2   Linear Equation Practice

1. Write the equation of the line with slope 3 passing through $(1, 4)$

   in slope-intercept form.

2. Graph the line with slope 3 which passes through $(1, 4)$ in the x-y plane.

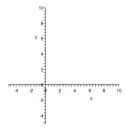

3 Domains of functions

Find the domains of the following functions and put your answers in interval notation.

1. $A(r) = \pi r^2$ is the area of a circle with radius r.

2. $h(t) = -16t^2 + 96$ is the height of a ball above ground level t seconds

         after being dropped from the top of a building.

3. $V(x) = (x)(16 - 2x)(20 - 2x)$ is the volume of a box formed from a piece of

         rectangular cardboard measuring 16 in by 20 in when squares of size $x \times x$

         sq. in are removed from the corners.

4 Math To Go

Draw a figure corresponding to the following word problem and carefully label any appropriate quantities with variables and/or numbers. Write down an equation to solve the problem. Solve the problem being sure to state your solution using the appropriate units.

1. The height of a lemon (in ft) t seconds after being thrown vertically upward is given by $h(t) = -16t^2 + 40t + 6$. The velocity of the lemon at time t seconds is given by $v(t) = -32t + 40$ ft/sec. At what time t is the velocity of the lemon 20 ft/sec?

Day 10

1   Daily Drill on Fundamentals: Factoring

Factor the following over the real numbers as much as possible.

1. $6x^2 + x - 2$

4. $x^3 - 27$

2. $x^2 - 5$

5. $x^3 + 2x^2 - 3x - 6$

3. $3x^2 + 7x - 6$

6. $12x^2 + 47x - 4$

2   Quadratic Equation Practice

1.   Solve $3x^2 - 6 = 0$

2.   Solve $9x^2 - 25 = 0$

3.   Solve $(x - 3)^2 - 81 = 0$

3   The Graph of Quadratic Equation: The Parobala

Find the x-intercepts of the following quadratic functions.

1. $f(x) = 2x^2 - 18$

2. $f(x) = 3x^2 - 2x - 5$

3. $f(x) = (x - 3)^2 - 2$

4. $f(x) = 5x^2 + 12$

5. $f(x) = x^2 - 10x + 25$

6. $f(x) = -(x - 2)^2 + 25$

4   Math To Go

Draw a figure corresponding to the following word problem and carefully label any appropriate quantities with variables and/or numbers. Write down an equation to solve the problem. Be sure to state your solution using the appropriate units.

1. The height of a lemon (in ft) t seconds after being launched vertically upward is given by $h(t) = -16t^2 + 12t + 10$. How long does it take for the lemon to hit the ground?

Day 11

1 Daily Drill on Fundamentals: Factoring

Factor the following over the real numbers as much as possible.

1. $6x^2 + 9x - 6$

4. $x^3 + 2$

2. $x^4 - x^3 - 6x^2$

5. $2x^4 - x^2 - 1$

3. $x^3 + 1$

6. $x^7 - 3x^5$

2 Drill on Domains

Find the domains of the following functions and put your answers in interval notation.

1. $\dfrac{3x^2 - 2x + 1}{x^2 - x - 20}$

2. $\dfrac{\sqrt{3x - 4}}{2x^2 - 11x + 9}$

3. $\dfrac{12x - 5}{x^2 + 24}$

3   The graph of a quadratic function

Draw and label the graph of the quadratic function indicated.

1. A parabola with vertex $= (3, 4)$ and with x-intercepts of 1 and 5.

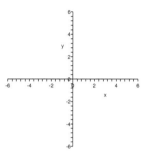

2. The graph of $f(x) = x^2 + 4x + 2$

4   Math To Go

Draw a figure corresponding to the following word problem and carefully label any appropriate quantities with variables and/or numbers. Write down an equation to solve the problem. Solve the problem being sure to state your solution using the appropriate units.

1. The height of a lime (in ft) t seconds after being tossed vertically upward is given by $h(t) = -16(t-2)^2 + 22$. What is the maximum height of the lime? At what time does the maximum height occur?

Day 12

1 Daily Drill on Fundamentals: Completing the Square

Complete the square on the following expressions (that is, write the expression in the form $a(x - h)^2 + k$ for appropriate values of a, h, and k).

1. $x^2 + 6x + 12$

2. $x^2 - 8x - 10$

3. $2x^2 - 4x - 12$

2 The Vertex of a Parabola

1.   Find the vertex of the parabola given by $y = x^2 - 10x + 32$.

2.   Find the vertex of the parabola given by $y = 3x^2 - 6x + 8$.

3.   Find the vertex of the parabola given by $y = x^2 - 3x + 5$.

3   Graphs of Quadratic Functions

Determine whether the given quadratic function takes on a maximum or minimum value.

1. $f(x) = -2x^2 - 14x + 6$

2. $f(x) = -(x - 4)^2 + 13$

3. $f(x) = 32 - 12x + 19x^2$

4   Math To Go

Draw a figure corresponding to the following word problem and carefully label any appropriate quantities with variables and/or numbers. Write down an equation to solve the problem. Be sure to state your solution using the appropriate units.

1. The height of a lemon (in ft) t seconds after being thrown vertically upward is given by $h(t) = -16t^2 + 40t + 6$. What is the maximum height of the lemon? At what time does the maximum height occur?

Day 13

1   Daily Drill on Fundamentals: Completing the Square

Complete the square on the following expressions (that is, write the expression in the form $a(x-h)^2+k$ for appropriate values of a, h, and k).

1. $x^2 + x + 1$

2. $2x^2 - 5x + 7$

3. $-5x^2 + 7x + 13$

2   The Vertex of a Parabola

1. Find the vertex of the parabola given by $y = x^2 - x - 1$.

2. Find the vertex of the parabola given by $y = -x^2 - 5x + 3$.

3. Find the vertex of the parabola given by $y = 3x^2 - 8x + 20$.

3   Maximum and Minimum Values of Quadratic Functions

Find the extreme (i.e., maximum or minimum) values of the given quadratic function as appropriate.

1. $f(x) = 3x^2 - 9x + 11$

2. $f(x) = -2x^2 - 10x + 15$

3. $f(x) = -x^2 + 12x - 36$

4   Math To Go

Draw a figure corresponding to the following word problem and carefully label any appropriate quantities with variables and/or numbers. Write down an equation to solve the problem. Solve the problem being sure to state your solution using the appropriate units.

1. The gas mileage, M, (in mpg) of a certain auto is modeled by $M(x) = -\frac{42}{(58)^2}(x - 58)^2 + 42$ where x is the speed of the auto (measured in mph). At what speed does the auto attain the best economical performance? What is this best gas mileage?

Day 14

1   Daily Drill on Fundamentals: Imaginary Numbers

Write the following in the form a + b*i* where *i* denotes $\sqrt{-1}$.

1. $\sqrt{-36}$

2. $\sqrt{-3}$

3. $3 - \sqrt{-5}$

4. $(2 + 3i)(-1 - 4i)$

5. $\dfrac{1 - i}{1 + i}$

6. $\dfrac{1}{3 - i}$

2   Quadratic Equation Practice

Solve the following equations in the complex numbers, $\mathbb{C}$.

1. $x^2 + x + 1 = 0$

2. $3x^2 - 2x + 1 = 0$

3. $-x^2 + 3x + 4 = 0$

3 The Discriminant: $b^2 - 4ac$

In the quadratic equation $ax^2 + bx + c = 0$ given below, use the *discriminant* $b^2 - 4ac$ to detect the number and type of solutions of the equation without solving.

1. $3x^2 - 4x - 12 = 0$

2. $-x^2 + 6x - 9 = 0$

3. $2x^2 + 4x + 8 = 0$

4 Math To Go

Draw a figure corresponding to the following word problem and carefully label any appropriate quantities with variables and/or numbers. Write down an equation to solve the problem. Be sure to state your solution using the appropriate units.

1. Suppose that the graph of a quadratic function, $y = f(x)$, has a vertex at $(3, -5)$ and that the equation $f(x) = 0$ has no real solutions. Does the corresponding parabola open upward or downward? Explain how you know.

Day 15

1   Daily Drill on Fundamentals: Domains of Functions

Find the domains of the following functions and put your answers in interval notation.

1. $3x^2 - 4x + 7$

4. $\dfrac{3x - 6}{2x^2 - 3x - 2}$

2. $\dfrac{x^2 + x - 20}{x - 4}$

5. $\dfrac{4x^2 - 1}{x^2 - 3x + 3}$

3. $\dfrac{x^2 - x - 6}{x^2 - x - 20}$

6. $\dfrac{3x + 1}{x^4 - x^2 - 12}$

2   Quadratic Equation Practice

Solve the following equations.

1. $3(x - 4)^2 - 5 = 0$

2. $4(x + 2)^2 - 9 = 0$

3. $2(3x - 7)^2 - 8 = 0$

3 A Parabola: The Graph of a Quadratic Function

Find the vertex of the corresponding parabola for each of the following quadratic functions.

1. $f(x) = -3(x-4)^2 + 8$

2. $f(x) = 2x^2 - 12x + 9$

3. $f(x) = -5x^2 + 18x - 1$

4 Math To Go

Draw a figure corresponding to the following word problem and carefully label any appropriate quantities with variables and/or numbers. Write down an equation to solve the problem. Solve the problem being sure to state your solution using the appropriate units.

1. Given the graph of $y = f(x) = 2x^2 - 12x + 10$ below, estimate the solutions to the inequality $2x^2 - 12x + 10 > 0$ and put your answer in interval notation. Then, find the solution algebraically.

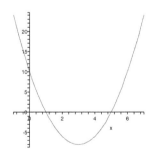

Day 16

1   Daily Drill on Fundamentals: Combining Functions

Let $f(x) = \sqrt{2x + 3}$ and $g(x) = \dfrac{x}{x - 1}$.

1.   Find $(f + g)(x)$ 4.   Find $(f \circ g)(x)$

2.   Find $(f \times g)(x)$ 5.   Find $(g \circ f)(x)$

3.   Find $\left(\dfrac{f}{g}\right)(x)$ 6.   Find $(g \circ g)(x)$

2   More composition

Let $f(x) = \dfrac{1}{x^2 + 3}$ and $g(x) = 5x^3 - 6$.

1.   Find $(f \circ g)(x)$

2.   Find $(g \circ f)(x)$

3 Decomposing Functions

Find functions f and g so that the given function h below can be written as $h = f \circ g$ where neither f nor g is the identity function. Note: Correct answers may vary.

1. $h(x) = 5x^3 - 9$

2. $h(x) = \sqrt{x - 7}$

3. $h(x) = \dfrac{3}{x^4 + 2}$

4. $h(x) = \dfrac{1}{\sqrt{x - 5}}$

4 Math To Go

1. Watches are on sale for 10% off. If sales tax is 8.25%, find the total price of a watch (to the nearest cent) which has a list price of $129. (Note: This calculation involves a sequential two-step process. That is, a composition of functions)

Day 17

1   Daily Drill on Fundamentals: Composition of Inverses

For f and g given below, find $(f \circ g)(x)$ and $(g \circ f)(x)$.

1. $f(x) = 3x + 1$

    $g(x) = \dfrac{x - 1}{3}$

3. $f(x) = \dfrac{1}{x}$

    $g(x) = \dfrac{1}{x}$

2. $f(x) = \sqrt[3]{\dfrac{x - 7}{5}}$

    $g(x) = 5x^3 + 7$

4. $f(x) = \dfrac{1}{x + 3}$

    $g(x) = \dfrac{1 - 3x}{x}$

2   Inverses

Find the inverses of the following functions if possible.

1. $f(x) = 5x - 8$

2. $f(x) = \sqrt[3]{\dfrac{6 - x}{2}}$

3 Inverse Meaning and Notation

Find the following. Assume the function f has an inverse (i.e., is *invertible*).

1. If $f(3) = 10$, find $f^{-1}(10)$.

3. If $f^{-1}(3) = 10$, find $f(10)$.

2. If $f(0) = 1$, find $f^{-1}(1)$.

4. If $f^{-1}(0) = 1$, find $f(1)$.

4 Math To Go

1. Recall that the graph of a function and it's inverse are reflections of each other through the line $y = x$. Use this feature to graph the inverse of the function $y = x^3$ given below.

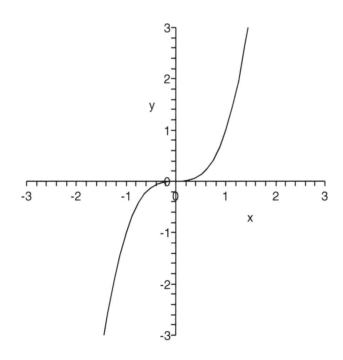

Day 18

1 Daily Drill on Fundamentals: Factoring (Exponential Function Emphasis)

Factor the following over $\mathbb{R}$ as much as possible.

1. $2^x + x2^x$

2. $(x+1)e^x + x^2 e^x$

3. $e^{2x} - e^x$

4. $e^{2x} - e^x - 6$

5. $x^3 3^x - 2x^2 3^x + x3^x$

6. $(x-2)^2 5^x - (x+2)^2 5^x$

2 Solving Equations

1.   Solve $2^x - 6 = 26$

2.   Solve $(x-3)3^x - 3^x = 0$

3.   Solve $(x^2 - 20)e^x - xe^x = 0$

3　More Factor Expansion

Expand the following by multiplying factors.

1.　　　$(5x - 2)(x - 3)$ 　　　　　　　　　　4.　　　$(4 - x)^3$

2.　　　$(2^x + 1)(2^x - 4)$ 　　　　　　　　5.　　　$(e^x + 1)^3$

3.　　　$(-x - 3)^2$ 　　　　　　　　　　　　6.　　　$(x - 1)^4$

4　Math To Go

1. The population of a culture of a bacterium is modeled by the equation $P(t) = P_0(1.012)^t$ where t is time measured in hours and P_0 is the initial count of bacteria. Is the population of bacteria increasing in number or decreasing. Explain how you know.

Day 19

1　Daily Drill on Fundamentals: Exponential vs. Logarithmic Notation

Write the following exponential equations in logarithmic form.

1.　　$3^4 = 81$　　　　　　　　　4.　　$3^L = 34$

2.　　$2^5 = 32$　　　　　　　　　5.　　$8^{N+4} = 102$

3.　　$\left(\frac{1}{2}\right)^3 = \frac{1}{8}$　　　　　　　6.　　$b^L = N$

2　Solving Equations

1.　　Solve $4^x = 16$

2.　　Solve $3^{x-2} = 27$

3.　　Solve $10^x = \frac{1}{100}$

3   More Form Conversion

Write the following exponential equations in logarithmic form.

1. $5^{2x} = 100$

4. $(4 - x)^3 = 1000$

2. $(x + 3)^4 = 19$

5. $(5x + 1)^{1/2} = y + 6$

3. $\left(\dfrac{1}{3}\right)^{x+1} = 12$

6. $\left(\dfrac{2}{3}\right)^{x+1} = y + 3$

4   Math To Go

1. A scientist wishes to determine the half-life, t, of a newly discovered isotope. After experimentation, she determines that the half-life (in seconds) is given by the solution to the equation $\frac{1}{32} = \left(\frac{1}{2}\right)^t$. What is the half-life of this isotope?

2. The doubling time (in minutes) of a virus population in a lab setting is determined to be the solution to the equation $256 = 2^t$. How many minutes does it take for the virus to double?

Day 20

1   Daily Drill on Fundamentals: Multiplying Algebraic Expressions

Expand the following by multiplying factors.

1. $(x)(x-3)(2x+6)$ 4. $(x^2)(x-30)(x^2+1)$

2. $(x^3)(x^2-x+1)$ 5. $(5x-4)(3x-2)$

3. $(3x+1)(x^3+5x)$ 6. $(x-1)^5$

2   Solving Algebraic Equations

1.   Solve $x^2-x-1=0$

2.   Solve $3x^3-2x^2=0$

3.   Solve $3x^2-4x-5=0$

3 The Degree and Leading Coefficient of a Polynomial

For each polynomial, identify the degree and the leading coefficient.

1. $3x^3 - 2x^2 + x - 1$

2. $-5x^2 + 7x + 2$

3. $3 - 4x + 6x^4$

4. $-\dfrac{1}{2}x^3 - 6x^2 + 2$

5. $17x^{99} - 48x^{30}$

6. 3

4 Math To Go

For each polynomial function given in factored form, identify the degree and the leading coefficient.

1. $f(x) = -3(x - 3)^2(3x + 1)^4 x^3(x + 2)^5$

2.

$g(x) = -3(1 - 2x)^3(x - 7)^4(x + 2)^5(x - 4)^9$

Day 21

1   Daily Drill on Fundamentals: Polynomial and Rational Functions

Find the domains of the following functions and put your answers in interval notation.

1. $f(x) = 3x^2 - 2x - 7$

4. $f(x) = \dfrac{x^2 - x - 6}{x^2 + x - 12}$

2. $f(x) = \dfrac{4x^3 - 6x}{2x - 7}$

5. $f(x) = \dfrac{1}{-6x^2 - 5}$

3. $f(x) = 3x - 2x^3 - x$

6. $f(x) = \dfrac{x^2 - x - 20}{2x^2 + 5x + 9}$

2   Solving Algebraic Equations

1.   Solve $\dfrac{1}{x} - \dfrac{3}{2x} = \dfrac{1}{5}$

2.   Solve $\dfrac{6x^2 + x - 2}{3x^2 - 6x - 8} = 0$

3.   Solve $\dfrac{x^3 + x^2 - 11x - 6}{x^3 - 6x} = 1$

3 Polynomial Features: End-Behavior

For each polynomial, determine the end behavior by filling in the appropriate symbol $+\infty$ or $-\infty$.

1. $f(x) = 3x^3 - 2x^2 + x - 1$

$f(x) \to$ ____ as $x \to \infty$

$f(x) \to$ ____ as $x \to -\infty$

3. $f(x) = 3 - 4x + 6x^4$

$f(x) \to$ ____ as $x \to \infty$

$f(x) \to$ ____ as $x \to -\infty$

2. $f(x) = -5x^2 + 7x + 2$

$f(x) \to$ ____ as $x \to \infty$

$f(x) \to$ ____ as $x \to -\infty$

4. $f(x) = -x^5 + 500x^4 + 2$

$f(x) \to$ ____ as $x \to \infty$

$f(x) \to$ ____ as $x \to -\infty$

4 Math To Go

Solve the following using polynomial division.

1. $x + 3 \overline{\smash{)}x^3 + 4x^2 + x - 6}$

Day 22

1 Daily Drill on Fundamentals: Even and Odd Functions

Determine whether the following functions are even, odd, or neither.

1. $f(x) = 2x^4 + 1$ 4. $f(x) = \sqrt{|x|}$

2. $f(x) = \sqrt[3]{x}$ 5. $f(x) = 5$

3. $f(x) = \dfrac{3x - 2}{x^3 - x}$ 6. $f(x) = \dfrac{1}{x}$

2 Solving Algebraic Equations

1.   Solve $\dfrac{2x^2 - 5x - 12}{x + 1} = 0$

2.   Solve $\dfrac{1}{x^2} = 0$

3.   Solve $\dfrac{1}{x^2} = 36$

3　Rational Functions: End-Behavior Again

For each rational function below, determine the end behavior by filling in the appropriate symbol: $+\infty$ or $-\infty$, or a real number.

1.　$f(x) = \dfrac{1}{x}$

$f(x) \to$ ____ as $x \to \infty$

$f(x) \to$ ____ as $x \to -\infty$

3.　$f(x) = \dfrac{x - x^5}{3 - x^5}$

$f(x) \to$ ____ as $x \to \infty$

$f(x) \to$ ____ as $x \to -\infty$

2.　$f(x) = \dfrac{3x^3 - 2x + 5}{-x^3 - 7x + 1}$

$f(x) \to$ ____ as $x \to \infty$

$f(x) \to$ ____ as $x \to -\infty$

4.　$f(x) = \dfrac{4x^3 - 2x + 7}{5x - 4}$

$f(x) \to$ ____ as $x \to \infty$

$f(x) \to$ ____ as $x \to -\infty$

4　Math To Go

Solve the following using polynomial division.

1.　$2x^3 + x \,\overline{)\,2x^5 - 2x^4 + x^3 - x^2}$

Day 23

1   Daily Drill on Fundamentals: Multiplying Algebraic Expressions

Expand the following by multiplying factors. Use rational exponents when appropriate.

1. $(3x - 1)(x^2 + x)$ 4. $(\sqrt{x})(4x + 1)(x^2 - 3)$

2. $(2x - 5)(2x + 5)$ 5. $(\sqrt{x} + 1)(2x + 7)$

3. $(x - 2)(x + 1)(x - 4)$ 6. $\sqrt[3]{(x + 2)^3}$

2   Solving Algebraic Equations

1.   Solve $3(x - 8) + 4(2x - 1) - 9 = 0$

2.   Solve $x^2 - x = 20$

3.   Solve $x^3 + 6x^2 + 12x + 8 = 0$

3 More Factor Expansion

Expand the following by multiplying factors. Use rational exponents when appropriate.

1. $(2x - 1)(x + 4)(x^3)$ 4. $(x - y)^3$

2. $(\sqrt{x} - 3y)(4y - 2\sqrt{x})$ 5. $(3x - y)^3$

3. $(x^2 - x)(4x + x^3)$ 6. $(x + y)^4$

4 Math To Go

Answer the following, then explain.

1. True or False: The solutions to $(x + y)(x - y) = 0$ are $x = y$ or $x = -y$. Explain your reasoning.

2. True or False: The solutions to $(x^2 + y^2)(x^2 - y^2) = 0$ are $x = y$ or $x = -y$. Explain your reasoning.

Day 24

1   Daily Drill on Fundamentals: Multiplying Algebraic Expressions

Expand the following by multiplying factors. Use rational exponents when appropriate.

1. $(3x)(4\sqrt{x} - \dfrac{2}{\sqrt{x}})$

4. $(x + y)(x - y)(x + z)$

2. $(x - y)^2(x^2 + y)$

5. $(2x - y)(3x - y)(z)$

3. $(2x - y)\left(\dfrac{x + y}{2}\right)$

6. $(3x - y)^2$

2   Solving Algebraic Equations

Solve the following equations over the real numbers. Use rational exponents when appropriate.

1. Solve $3x^2 - 6 = 21$

2. Solve $(\sqrt{x} - 3)(3\sqrt{x} + 5) = 0$

3. Solve $(x^2 - 16)(x + 16)\left(\dfrac{4x - 3}{3}\right) = 0$

3   More Factor Expansion

Expand the following by multiplying factors.

1. $(\sqrt[3]{x} - 1)(\sqrt[3]{x} - 2)(\sqrt[3]{x} - 3)$ 4. $(u + v)^4$

2. $(x + y)^2(x - y)$ 5. $(x + y)^5$

3. $(u - 2v)^3$ 6. $(y - x)^4$

4   Math To Go

Draw a figure corresponding to the following word problem and carefully label any appropriate quantities with variables and/or numbers. Write down an equation to solve the problem. Be sure to state your solution using the appropriate units.

A rancher wishes to build a corral of area 12,600 square feet. He will use 500 feet of fencing to create this corral. Let x denote the length of the corral and y denote its width.

1. Write an equation in x and y that denotes the desired area of the corral.

2. Write an expression in x and y that denotes the total fencing that will be used to make the corral.

3. Find the dimensions of the corral if the farmer uses all the fencing available.

Day 25

1   Daily Drill on Fundamentals: Algebraic Expressions

Solve for the given variable in the following formulas. Assume all variables denote positive quantities.

1. h in $V = \dfrac{1}{3}\pi r^2 h$

4. i in $A = P(1+i)^n$

2. l in $P = 2l + 2w$

5. n in $A = P(1+i)^n$

3. h in $A = \pi r \sqrt{r^2 + h^2}$

6. P in $V = P\left(\dfrac{(1+i)^n - 1}{i}\right)$

2   Evaluating Expressions using a Calculator

Find the value indicated. Round to two decimal places as necessary.

1.   Find A in $A = Pe^{rt}$ if $P = 500, r = .065, t = 4$.

2.   Find A in $A = P\left(\dfrac{(1+i)^n - 1}{i}\right)$ if $P = 800, i = .0055, n = 240$.

3.   Find A in $A = P(1+i)^n$ if $P = 250, i = .00875, n = 60$.

3 Basic Formulas

Find the following. Be sure to include the appropriate units.

1. The area of a right triangle with leg lengths 9 and 12 cm.

2. The perimeter of a rectangular whose length is 8 cm and whose width is 6 cm.

3. The area enclosed by a circle whose radius is $\sqrt{5}$ cm.

4. The area of a trapezoid with base lengths 4 cm and 9 cm and height 7 cm.

5. The area of a right triangle with hypotenuse length 13 cm and a leg length of 12 cm.

6. The radius of the circle (in cm) whose circumference is half the enclosed area.

4 Math To Go

Draw a figure corresponding to the following word problem and carefully label any appropriate quantities with variables and/or numbers. Write down an equation to solve the problem. Be sure to state your solution using the appropriate units.

1. A conical pile of sand has slant height of 10 ft. If its base radius is 6 ft, find the amount of sand in the pile. **Hint:** The volume of a cone is given by $V = \frac{1}{3}\pi r^2 h$.

Day 26

1 Daily Drill on Fundamentals: Factoring (Negative Exponent Emphasis)

Factor the following to simplify. Write your answer without using negative exponents.

1. $a^3 b^{-5} + a^{-1} b^2$

4. $x^{-3} y^5 - \dfrac{x}{y}$

2. $x^{-2} y^{-3} + x^{-3} y^{-4}$

5. $a^{-3} b^3 - a^2 b^{-4} + a^{-5} b^{-2}$

3. $x^{-3} \sqrt{2x-1} - x\sqrt{2x-1}$

6. $(x^2 - x - 1)^3 (5x+3)^{-2} + (x^2 - x - 1)^{-3} (5x+3)^{-6}$

2 Linear Equation Practice

1.   Find the slope of the line passing through $(3, 2)$ and $(7, -22)$.

2.   Find the equation of the line passing through $(3, -1)$ which is parallel to $y = -4x + 5$.

3.   Find the y-intercept of the line passing through $(5, 4)$ and $(2, -2)$.

3   Combining and Simplifying Expressions

Combine and simplify the following expressions.

1. $\dfrac{1}{x^2} + \dfrac{2}{x}$

4. $\dfrac{1}{\sqrt{x-7}} + \dfrac{x}{x-7}$

2. $\dfrac{3x-2}{x-4} + \dfrac{x-1}{x+2}$

5. $\dfrac{3xy^2}{x^2y^3} - \dfrac{x}{xy^5}$

3. $\dfrac{x-4}{x^2-x-20} - \dfrac{2x}{x-5}$

6. $x^3y^{-4} + x^{-1}y^2$

4   Math To Go

1. Eric and George have a disagreement about their math homework. Eric says the answer is

   $$3x^2(2x^2 - 9x)^{-1} - (4x^4 - 9x^3)(2x^2 - 9x)^{-2}$$

   while George claims the answer is

   $$\frac{2x^4 - 18x^3}{(2x^2 - 9x)^2}.$$

   Show that both answers are, indeed, identical to each other.

Day 27

1   Daily Drill on Fundamentals: Exponential vs. Logarithmic Notation

Write the following logarithmic equations in exponential form.

1. $\log_2 16 = 4$

4. $\ln 1 = 0$

2. $\log_5 \dfrac{1}{125} = -3$

5. $\log_b N = L$

3. $\log 1000 = 3$

6. $\log_{\frac{1}{3}} 9 = -2$

2   Solving Inequalities

Put your answers in interval notation below.

1.   Solve $3x - 7 \le 11$

2.   Solve $x^2 - 6x - 7 > 0$

3.   Solve $\dfrac{x^2 - x - 30}{x^3 - 1} \le 0$

3   More Form Conversion

Solve without a calculator.

1. $\log_3 27$ 4. $\ln e^3$

2. $\log \dfrac{1}{10}$ 5. $\log_{\frac{1}{2}} 8$

3. $\log_6 36$ 6. $\log_{\frac{1}{5}} 1$

4   Math To Go

1. The Richter scale is used to measure the intensity of earthquakes as compared to a "standard" earthquake. It is based on the base-ten logarithm. The magnitude, R, of the quake is computed as $R = \log I$ where I is the intensity of the quake. Thus, $I = 10^R$. As an example, an earthquake that measures 6.2 on the Richter scale can be seen to be about 31.6 times as strong in intensity as an earthquake measuring 4.7 on the Richter scale by computing

$$\frac{10^{6.2}}{10^{4.7}} \approx 31.6.$$

The San Francisco earthquake of 1906 measured 7.8 on the Richter scale. The quake and resulting fires destroyed about 80% of the city and killed approximately 3000 people. The largest recorded eathquake occurred in Valdivia, Chile in 1960. It measured 9.6 on the Richter scale. About how much stronger in intensity was the Valdivia quake as compared to the San Francisco quake?

Day 28

1   Daily Drill on Fundamentals: Multiplying Algebraic Expressions

Expand the following by multiplying factors. Write your answers using rational exponents.

1. $x^{\frac{3}{2}}(x^2 - x^{-1})$

4. $x^2 y^{\frac{2}{3}}(xy - x^2 y^{\frac{1}{3}})$

2. $\sqrt{3x - 4} \cdot \sqrt[3]{3x - 4}$

5. $(x^{\frac{1}{2}} + 1)^2$

3. $(x^{\frac{1}{2}} - 3)(x^{-\frac{2}{3}} + 1)$

6. $(x^{\frac{1}{3}} - 1)^3$

2   Solving Algebraic Equations

1.   Solve $\sqrt[3]{3x} = 2$

2.   Solve $(x - 3)^{\frac{1}{3}}(3x + 5)^{-1} = 0$

3.   Solve $\sqrt{x(x - 7)}\left(\dfrac{x - 1}{3}\right) = 0$

M.J. Sanders

3   Systems of Equations

Solve the following systems of equations.

1. $2x + y = 18$

          $-x - y = -13$

2. $3x - 2y = 11$

          $4x + y = 11$

3. $5x - 4y = -14$

          $3x - 5y = -11$

4   Math To Go

To convert temperature measured in degrees Fahrenheit, F to temperature in degrees Celsius, C, we apply the conversion

$$C^\circ = \frac{5}{9}(F^\circ - 32)$$

1. There is one temperature, x, at which $x\,^\circ F = x\,^\circ C$. What is this temperature?

Day 29

1   Daily Drill on Fundamentals: Domains

Find the domains of the following functions and put your answers in interval notation.

1. $f(x) = \dfrac{2x - 1}{x^2 + 8}$

4. $f(x) = \log(x - 4)$

2. $f(x) = \dfrac{\sqrt{x - 1}}{x - 10}$

5. $f(x) = \dfrac{x}{e^x}$

3. $f(x) = \ln x$

6. $f(x) = \dfrac{\sqrt[3]{x}}{e^x - 1}$

2   Solving Inequalities

Solve the following and put your answers in interval notation.

1.   Solve $\dfrac{2x - 8}{3} \geq 5$

2.   Solve $x^2 \leq x$

3.   Solve $\dfrac{x^2 - x + 1}{e^x} > 0$

3   More End-Behavior

For each function, determine the end behavior by filling in the appropriate symbol $+\infty$ or $-\infty$ or a real number.

1. $f(x) = \dfrac{-12x^3 + x^2}{5}$

   $f(x) \to$ ____ as $x \to \infty$

   $f(x) \to$ ____ as $x \to -\infty$

3. $f(x) = \dfrac{-2x^2 - x + 7}{3x^3 + 6x}$

   $f(x) \to$ ____ as $x \to \infty$

   $f(x) \to$ ____ as $x \to -\infty$

2. $f(x) = e^x + 4$

   $f(x) \to$ ____ as $x \to \infty$

   $f(x) \to$ ____ as $x \to -\infty$

4. $f(x) = \dfrac{4x^5 - 2x^2 + x + 1}{-x^3 + x^2 - x + 9}$

   $f(x) \to$ ____ as $x \to \infty$

   $f(x) \to$ ____ as $x \to -\infty$

4   Math To Go

Solve the following using polynomial division.

1. $3x^2 + x \,\overline{)\,3x^4 - 5x^3 + x^2 + x}$

Day 30

1   Daily Drill on Fundamentals: Interval Notation

Write the following sets of real numbers using interval notation.

1. $\{x : 3 < x \leq 10\}$ 4. $\{x : 3 > x \geq -2\}$

2. $\{x : x > -2\}$ 5. $\{x : |x| \neq 3\}$

3. $\{x : x \leq 0 \text{ or } x > 6$ 6. $\{x : |x + 2| > 5\}$

2   Equations of Lines

Put your answer in slope-intercept form below.

1.   Find the equation of the line passing through $(2, 3)$ and $(5, -6)$.

2.   Find the equation of the line passing through $(2, 3)$ which is parallel to the line given by $y - 4 = 5(x - 3)$.

3.   Find the equation of the line passing through $(-1, -5)$ which is perpendicular to the line given by $10y - 5x = 14$.

3 Factoring

Factor the following over the real numbers as much as possible.

1. $x^2 + 7x + 12$

4. $x^3 - 3x^2 + 3x - 1$

2. $x^5 - 81x^3$

5. $x^3 - 1$

3. $x^2 - x + 1$

6. $xe^x - 3xe^{2x}$

4 Math To Go

1. Kim scored 88, 84, and 94 on the first three exams in her math course. If each of these exams is worth 20% of her course grade and the final exam is worth 40%, how high must she score on her final to receive an A (90% or better) in the course?

Congratulations!!! You've completed 30 days worth of readiness exercises!

PART II: Daily Work Notes

Day 1 Work Notes:

Expanding or multiplying factors utilizes the distributive property of multiplication over addition as its basis. For example, to find

$$(x^2 + x + 1) \cdot (2x - 3)$$

every term of the first factor must be multiplied by every term of the second factor. Observe:

$$(x^2 + x + 1) \cdot (2x - 3) = x^2 \cdot 2x + x^2 \cdot (-3) + x \cdot 2x + x \cdot (-3) + 1 \cdot 2x + 1 \cdot (-3) = 2x^3 - 3x^2 + 2x^2 - 3x + 2x - 3.$$

Simplifying,

$$(x^2 + x + 1) \cdot (2x - 3) = 2x^3 - x^2 - x - 3.$$

To *solve an equation* means to find all the values of all variables that make the equation true. One basic tool for doing such is the "Zero-Product Property". It states that if two quantities are multiplied together and their resulting product is zero, then one of them had to be zero to begin with. For example, to solve

$$(x - 3) \cdot (x + 1) = 0,$$

we use this property to say $x - 3$ must be zero or $x + 1$ must be zero. Thus we arrive at the solutions to the equation,

$$x = 3 \text{ or } x = -1.$$

Pascal's triangle is a useful way for identifying the coefficients in multiplying a binomial by itself a number of times. The first few rows of Pascal's triangle are illustrated below.

```
                    1
                1       1
            1       2       1
        1       3       3       1
    1       4       6       4       1
1       5      10      10       5       1
.
.
.
```

Each line of the triangle begins and ends with a 1. The interior terms of each line can be computed as the sum of the two "diagonal" terms immediately above that term. The coefficients of the terms of the expansion of $(a + b)^n$ are found in the $(n + 1)^{st}$ row of Pascal's triangle. The powers of the "a-terms" descend while the powers of the "b-terms" ascend. For example, utilizing the fourth row of Pascal's triangle, we have,

$$(a + b)^3 = 1a^3 + 3a^2b + 3ab^2 + 1b^3, \text{ or more simply, } a^3 + 3a^2b + 3ab^2 + b^3.$$

While this is not the only way of determining the coefficients, the ease of computing the first few rows of the triangle provides a convenient mechansim for low-degree factor expansion.

Note that if it is a difference such as $(a - b)^3$, the signs alternate beginning with a positive first term. That is,

$$(a - b)^3 = a^3 - 3a^2b + 3ab^2 - b^3.$$

Day 2 Work Notes:

Word problems, or applications, are notorious for being difficult. In fact, they are difficult for mathematicians and non-mathematicians alike. This is because there is extra work involved in understanding the problem from a word description and transforming it into some mathematical object to solve (such as an equation or an inequality). Given a word problem, read the problem carefully several times until you can begin to grasp the issue. One detail to focus on is to carefully identify what is being asked for. That is, what will it take to solve this problem. Drawing a picture or schematic to represent the problem is frequently a valuable tool. One aspect that can be confusing is that you, the solver, are often free to identify variables with whatever choice of letters or symbols you choose. This is good as your choice of variable names may help you to keep track of information. For example, you might choose l to denote the length of a rectangle and w to denote it's width. Someone else might choose b for the base and h for the height of the same rectangle. These are both fine choices. Don't be afraid to own the problem by changing letters if it seems confusing. The best tactic for becoming a good problem solver is to do lots of problems. It will get easier!

Day 3 Work Notes:

Factoring involves splitting a mathematical expression into pieces. It is the opposite of expanding or multiplying expressions. Frequently we work with polynomial expressions like you see in the exercises for today.

Factoring is an incredibly important skill for doing mathematics. It is used in solving equations, simplifying expressions, and a whole other host of applications. One thing to keep in mind is that we are usually interested in factoring over the real numbers (versus over the larger set of complex numbers- more on this later). Regarding polynomials, it is a mathematical fact that every polynomial can be factored into linear factors and what are called *irreducible quadratic* factors. Linear factors are expressions that can be written in the form $ax + b$ for $a, b \in \mathbb{R}$ (that is, for a and b in the real numbers). Irreducible quadratic factors are harder to identify. You will work more with quadratics in the exercises that follow. For now, given a quadratic factor

$$ax^2 + bx + c,$$

the quadratic factor is an irreducible quadratic factor, or in other language, is irreducible over $\mathbb{R}$, provided that the quantity $b^2 - 4ac$ is negative. This quantity is called the *discriminant* of the quadratic expression. The condition is equivalent to saying that the quadratic equation

$$ax^2 + bx + c = 0$$

has no real solutions. Some quadratics are easy to identify using this latter criterion: For example, $x^2 + 1$ is irreducible since x^2 is never negative and adding 1 makes it positive. Thus, there are no real solutions to the equation $x^2 + 1 = 0$ which means that it is irreducible. Similarly, the quadratic $x^2 + c$ for any $c > 0$ is, like-wise, irreducible. Other quadratic factors require a little more observation to detect. For example, $x^2 + x + 1$ is also irreducible as you can check with the discriminant.

Becoming proficient at factoring requires practice and effort. Several formulas that are useful for beginning to factor involve differences of squares (i.e., expressions of the form $a^2 - b^2$) as well as sums and differences of cubes.

Difference of Squares:	$a^2 - b^2 = (a + b)(a - b)$
Sum of Cubes:	$a^3 + b^3 = (a + b)(a^2 - ab + b^2)$
Difference of Cubes:	$a^3 - b^3 = (a - b)(a^2 + ab + b^2)$

Note that the sum of squares is not listed above. It cannot, in general, be factored further over the real numbers.

The *slope* of a line in the x-y plane is a measure of the line's steepness. It is frequently (but not always) denoted by m. Given a line passing through the points (x_1, y_1) and (x_2, y_2) in the plane, the slope of the line is defined by

$$\text{slope} = m = \frac{y_2 - y_1}{x_2 - x_1} \text{ provided } x_2 - x_1 \neq 0.$$

It should be pointed out that a vertical line has no defined slope (this is precisely the case when $x_1 = x_2$ or, equivalently when $x_2 - x_1 = 0$ giving zero in the denominator of the formula above. Division by zero is undefined).

The equation of a line can take on different forms. The equation of a line with slope m and passing through the point (x_1, y_1) can be written as

$$y - y_1 = m(x - x_1).$$

This is called the *point-slope form* of the equation of a line. For example, the point-slope form of the equation of the line with slope $m = 3$ and passing through $(x_1, y_1) = (-2, 5)$ is

$$y - 5 = 3(x - (-2)) \implies y - 5 = 3(x + 2).$$

The above form is useful for writing the equation of a line when its slope is given and a known point lies on the line.

Solving the above equation for y in terms of x, we get

$$y = 3x + 11.$$

This equation is said to be in *slope-intercept form*. Notice that the slope is readily apparent as the coefficient of x. The "11" is the *y-intercept* of the line. That is, the point $(0, 11)$ on the y-axis lies on the graph of the line. In general, equations of the form

$$y = mx + b$$

are said to be in slope-intercept form where $m, b \in \mathbb{R}$. In the slope-intercept form of the equation of a line, the line can be graphed easily. For example, we know the point $(0, 11)$ lies on the line. If we let x be any number other than 0, we obtain a new point on the line. For example, the point $(1, 14)$ also lies on the line. Since a line is completely determined by connecting two points on the line with a straight-edge, the graph can be produced quickly.

It is noted that the *general form* of the equation of a line is an equation written in the form

$$Ax + By + C = 0.$$

There is not much mathematical use for this form except to say that every line has an equation that can be put in this form (even vertical lines) and every equation of this form represents a line (provided A and B are not both zero).

The standard for writing sets of real numbers in college mathematics is to use interval notation. The following are examples of 1) a closed interval, 2) an open interval, 3) a half-open interval. While there is more to be said, some examples of the meanings of the intervals given on the left are provided using *set-builder notation* on the right.

1) $[-10, 10] = \{x \in \mathbb{R} : -10 \leq x \leq 10\}$

2) $(-5, 5) = \{x \in \mathbb{R} : -5 < x < 5\}$

3) $[-1, 1) = \{x \in \mathbb{R} : -1 \leq x < 1\}$

For *unbounded intervals*, we use the symbols $-\infty$ or ∞. As examples,

$$(-\infty, -10) = \{x \in \mathbb{R} : x < -10\}$$

$$[10, \infty) = \{x \in \mathbb{R} : x \geq 10\}$$

It should be observed that $\pm\infty$ should always be paired with parentheses when using interval notation in introductory college math courses. This is because $\pm\infty$ do not represent numbers, but, instead, they represent ideas. Indeed, they both represent the notion that the real numbers are unbounded; that is, the real numbers go on in both directions forever and ever. As a final example for now,

$(-\infty, \infty)$ denotes the entire set of real numbers,

$$(-\infty, \infty) = \mathbb{R}.$$

Day 4 Work Notes:

As observed in yesterday's exercise set, factorization formulas are useful tools when working with polynomials or other expressions. While there are certainly generalizations of the formulas provided below, recall several basic formulas now:

Difference of Squares:	$a^2 - b^2 = (a + b)(a - b)$
Sum of Cubes:	$a^3 + b^3 = (a + b)(a^2 - ab + b^2)$
Difference of Cubes:	$a^3 - b^3 = (a - b)(a^2 + ab + b^2)$

Also recall that the sum of squares cannot, in general, be factored further over the real numbers.

Pattern recognition is a necessary skill in mathematics. It does not always come readily. As an example, consider the expression $x^3 + 8$. If we recognize 8 as a perfect cube ($8 = 2^3$) we have the expression $x^3 + 8 = x^3 + 2^3$ which can be factored as

$$x^3 + 8 = x^3 + 2^3 = (x + 2)(x^2 + 2x + 4).$$

As you can check, $x^2 + 2x + 4$ is irreducible. Thus, the above is the complete factorization over $\mathbb{R}$. Moreover, in a similar fashion, since $5 = (\sqrt[3]{5})^3$, observe that

$$x^3 + 5 = (x^3 + (\sqrt[3]{5})^3) = (x + \sqrt[3]{5})(x^2 + \sqrt[3]{5}x + \sqrt[3]{5^2}).$$

Fitting expressions to given patterns takes practice too.

Day 5 Work Notes:

A linear equation is an equation that can be written in the form

$$ax + b = 0 \text{ where } a \neq 0.$$

Linear equations in this form are straight-forward to solve: Subtract b from both sides of the equation and then divide by a.

Thus, the solution to $ax + b = 0$ is given by $x = \dfrac{-b}{2a}$

However, not all linear equations come given in this standard form. For example, $2(x - 3) + 5x = -4$ is also a linear equation. If our goal is to solve this equation, the best strategy is to move to isolate x on one side of the equation and observe the constant on the other side. There is no need to write it in the form $ax + b = 0$. Indeed, doing so usually requires more work than is necessary. To solve the above equation, we perform

$$2(x - 3) + 5x = 4 = 2x - 6 + 5x = 4 \implies 7x = -10$$

so that $x = -\dfrac{10}{7}$ is the solution to the equation.

Day 6 Work Notes:

We frequently speak of *rational numbers* in mathematics or simply say that a number is *rational*. It is important to understand the significance of this terminology. The set of rational numbers, sometimes denoted by $\mathbb{Q}$, can loosely be described as the set of fractions. This is not technically correct though because any number can be written as a fraction. For example,

$$\pi = \frac{\pi}{1}$$

More precisely, the set of rational numbers is defined by

$$\left\{ \frac{a}{b} : a \text{ and } b \text{ are integers with } b \neq 0. \right\}$$

Recall that the integers are the set of all positive and negative whole numbers. It is not obvious that there are any numbers that are not rational (called *irrational numbers*) but, indeed, there are infinitely, infinitely many. Some "famous" irrational numbers, are $\{\pi, e, \sqrt{2}, \text{ etc.}\}$. Today, we are interested in rational exponents and their meaning. Recall that the square root of a non-negative number x (sometimes referred to as the principal square root of x) is denoted with a radical symbol, $\sqrt{x}$, and is defined to be the non-negative number y so that $y^2 = x$. We also encounter cube roots, fourth roots, etc. In general, the nth root of x is defined by

$$\sqrt[n]{x} = y \text{ where } y^n = x \text{ and } n > 1 \text{ is a positive integer.}$$

If n is even, we require x, y to be non-negative in order to remain within the real number system. We frequently have the desire to work with rational exponents in mathematical calculations. To do this, we define

$$a^{\frac{1}{n}} = \sqrt[n]{a} \text{ for } n \text{ a positive integer greater than 1.}$$

Again, a is not allowed be negative when n is even to remain in the real number system.

More generally yet, we define

$$a^{\frac{m}{n}} = \sqrt[n]{a^m}$$

for appropriate values of m and n. We will only use these rational exponents provided

$$\sqrt[n]{a^m} = (\sqrt[n]{a})^m.$$

Equivalently, this is the same thing as saying that our rational exponents will always be written in reduced form. Thus,

$$a^{\frac{m}{n}} = \sqrt[n]{a^m} = (\sqrt[n]{a})^m.$$

As examples, we have

$$36^{\frac{1}{2}} = \sqrt{36} = 6, 8^{\frac{2}{3}} = \sqrt[3]{8^2} = (\sqrt[3]{8})^2 = 4, \text{ and } (-32)^{\frac{3}{5}} = \sqrt[5]{(-32)^3} = (\sqrt[5]{-32})^3 = -8.$$

Day 7 Work Notes:

Some equations are said to be *linear-like*. While not linear, the idea is that through manipulation, we are able to transform the equation into a linear one and then solve the new equation. There are dangers that can arise. For example, extraneous solutions (numbers that appear to be solutions but which are not) can appear. One place where linear-like equations can appear involves equations with fractions. A good strategy for solving such equations is to multiply through the equation by the least common denominator of all fractions involved. This removes the fractions and can result in various types of equations. Here, we'll see linear equations. For example, consider,

$$\frac{1}{x-1} + \frac{2}{x} = \frac{4}{x^2 - x}.$$

Multiplying through the equation by $(x-1)(x)$, we have

$$x + 2(x - 1) = 4$$

and solving this linear equation gives $x = 2$ which is an actual solution as you can check in the original equation. If we would have obtained $x = 0$ or $x = 1$, these "solutions" would have been extraneous; they would not have satisfied the original equation. Extraneous solutions may arise because multiplying through an equation by a variable quantity can result in (unknowingly) multiplying by zero. That operation is not allowed in generating equivalent equations. If this seems confusing, just remember to check your answers in the original equation. This is the reason Exercise 6 in today's exercise set has no solutions.

Functions are main objects of study in mathematics. They provide rules for specifying operations on numbers. The variable of a function is of little importance. For example, the function $f(t) = t^2 + 1$ is identical to the function $f(x) = x^2 + 1$. It is the rule or procedure that is important. The rule says, given an input value, square it and add one to get the output value.

There is a convention for functions. Given an algebraic function such as

$$f(x) = \frac{2x + 1}{x - 3},$$

the *domain* of the function (or, set of all allowable inputs) is implied to be the set of all real numbers x so that $f(x)$ is a real number. You are expected to know this convention without it being stated and to be able to find the implied domain. Since the function $f(x)$ mentioned above does not make sense when $x = 3$, but provides a real number output for all other real numbers, the domain of the function f is

$$\{x \in \mathbb{R} : x \neq 3\}.$$

Or, in interval notation, the domain is $(-\infty, 3) \cup (3, \infty)$.

In general, fractions cause problems when the denominator is zero. Also, recall that even roots of negative numbers are not real. Thus, you must account for that condition when determing domains as well.

As you will see in several days, occasionally the context of a function can determine the domain too. For example, if $A(r) = \pi r^2$ denotes the area of a circle of radius r, then the domain of A is $\{r \in \mathbb{R} : r > 0\}$ even though the formula is defined for all real numbers r. This is because the radius of a circle must be positive. Thus, domains can be affected by context in applied problems as well.

Day 8 Work Notes:

Equivalent algebraic expressions are those which give the same outputs for the same input values. The expressions must be defined for exactly the same set of numbers. If we regard these expressions as functions, we could say that two expressions are equivalent if they provide the same rule for obtaining the same outputs and have the same domain. For example, the expressions

$$\frac{3x - x^2}{x} \text{ and } \frac{6x - 2x^2}{2x}$$

are equivalent expressions while

$$1 \text{ and } \frac{x}{x} \text{ are not.}$$

Day 9 Work Notes:

The graph of a function $y = f(x)$ in the x-y plane is the set of all ordered pairs $(x, f(x))$ where x is in the domain of f. Graphs of functions give us a picture of the function and are very useful for detecting behavior of the function that is difficult to see simply by observing the function's equation or formula.

Perhaps the easist functions to graph are linear functions. Their graphs are lines. To graph a line, once we know two points that lie on the line, we connect them with a straight edge and we're done. Other functions require more effort. One old-fashioned approach is to graph by the point-plotting method. Simply choose a handful of convenient input values, plot the corresponding ordered pairs $(x, f(x))$ and connect with a smooth curve. This procedure is rather crude and can be full of pitfalls. Indeed, calculators and computers graph in the same fashion; only, they use thousands of ordered pairs. These devices can still encounter the same errorful pitfalls; namely, that connecting dots with a smooth curve can miss important behavior. It is worthwhile to develop skills in graphing basic types of functions accurately: power functions, root functions, etc. We'll work more with graphs of quadratic functions tomorrow.

Recall that a function's domain in an applied setting can sometimes be determined by the physical constraints of the problem. For example, if $P(t)$ denotes the population of a bacterium t minutes after being placed in a petri dish, then we are only interested in $t \geq 0$ as "negative time" doesn't make sense. Thus, the domain of $P(t)$ would be $[0, \infty)$ in this example.

You will encounter exercises in today's set whose intent is to get you to think about these matters.

Day 10 Work Notes:

Quadratic functions are functions that can be written in the form $f(x) = ax^2 + bx + c$ where $a \neq 0$. Of course, quadratic functions may not all come written in this form. For example,

$$f(x) = 2(x-1)^2 + 5$$

is a quadratic function too as can be seen by expanding and removing the parentheses. In fact, this form of a quadratic equation is exceptionally nice and is called the *standard form* of a quadratic equation.

Today, you will be solving quadratic equations. We are only interested in finding real solutions for now. It is important to recognize that a quadratic equation may not have any real solutions. Or, it may have exactly one real solution or it may have two distinct real solutions. These are the only possibilities however. For equations that come in the form

$$a(x-h)^2 + k = 0$$

for fixed numbers a, h, k, this process is quite straight forward. (The choice of letters here is standard, more will be said later.)

For example, suppose we wish to solve the equation

$$3(x-2)^2 - 48 = 0.$$

To understand how to proceed, let's consider a nicer problem. Namely, solving $x^2 = 15$. As you likely know, we'd like to take square roots of both sides. There is frequently confusion that arises on this topic. The key step here is to understand that $\sqrt{x^2}$ is the absolute value of x. *That* is what induces the $\pm$ symbol you may be expecting to arise. Consider the following sequence:

$$x^2 = 15$$
$$\sqrt{x^2} = \sqrt{15}$$
$$|x| = \sqrt{15}.$$

Now if the absolute value of x is $\sqrt{15}$, then x must be $\pm\sqrt{15}$. Almost everyone performs this last step automatically without any consideration of $|x|$. And, that's fine. Thus, a more common work sequence looks like:

$$x^2 = 15$$
$$\sqrt{x^2} = \sqrt{15}$$
$$x = \pm\sqrt{15}.$$

Now, back to our original problem of solving $3(x-2)^2 - 48 = 0$, our strategy will be to isolate the square on one side of the equation and then take square roots of both sides using the $\pm$ shortcut as above. Observe:

$$3(x-2)^2 - 48 = 0$$
$$3(x-2)^2 = 48$$
$$(x-2)^2 = 16$$
$$x - 2 = \pm\sqrt{16} = \pm4.$$

Solving for x, we get,

$$x = 2 \pm 4 = \{-2, 6\} \text{ as solutions.}$$

This is a method worth remembering. You'll practice it today. Note that we obtain two real solutions because the quantity on the right above is positive. If it were zero, we would have one real solution and if it were negative, we would have no real solutions.

Regarding the graph of a quadratic equation, recall that the graph is called a *parabola*. The "turning point" is called the *vertex* of the parabola. While we'll work with these notions more in the next several days, for today's exercises we need to understand that the graph of a quadratic function $y = f(x) = ax^2 + bx + c$ has precisely zero, one, or two x-intercepts (i.e., points where the graph crosses the x-axis). These correspond, respectively, to the situations where the the quadratic equation $ax^2 + bx + c = 0$ has zero, one, or two real solutions (after all, the x-intercepts occur on the graph where the y-coordinate ia zero).

Day 11 Work Notes:

If we know the vertex (h, k) of a quadratic function, it is easy to graph. You may either plot the x-intercepts if they're easy to find (and exist!) or simply plot the vertex and choose values of x other than h and use the point-plotting method. Remember that graphs of quadratic functions are symmetric about the vertical line $x = h$. In the *standard form* of a quadratic function, that is, the form $y = a(x - h)^2 + k$, the vertex is easily observed: It is the point (h, k). By the method of "completing the square", something we'll practice tomorrow, it can be shown that the vertex of the parabola given by $y = f(x) = ax^2 + bx + c$ is given by

$$\text{Vertex } = \left(-\frac{b}{2a}, f\left(-\frac{b}{2a} \right) \right).$$

Since the vertex of a parbola represents the turning point of the graph, optimization problems (i.e., finding a maximum or minimum output values or where they occur) involving quadratic functions hinge on identifying the coordinates of the vertex.

Day 12 Work Notes:

Completing the square is an important operation in mathematics. It arises in multiple places and is a skill that is needed to be retained. For our work today, we are interested in transforming a quadratic expression of the form $ax^2 + bx + c$ into the form $a(x - h)^2 + k$. We've seen that this second form is quite useful. The power of this operation is of great use in other places as well. To begin, we'll point out that the only expressions that we ever complete the square on, are expressions of the form

$$x^2 + bx.$$

Notice that the coefficient of x^2 is 1 here. This is important. We'll examine a more general situation later. To complete the square when the leading coefficient is 1, we follow the following steps:

1. Take half the coefficient of x, square it and add it in.

2. To preserve the original expression, we subtract the same quantity symbolically.

3. Observe the perfect square and identify it as a square.

The following example illustrates. We'll complete the square on $x^2 + 10x$:

Given $x^2 + 10x$, we take half the coefficient of x, square it and add it in resulting in

$$x^2 + 10x + (5)^2$$

We then subtract the same quantity symbolically, obtaining

$$x^2 + 10x + (5^2) - 25$$

Observing that the first three terms form a complete square, we write

$$x^2 + 10x = (x^2 + 10x + (5)^2) - 25 = (x+5)^2 - 25$$

Thus, we have $x^2 + 10x = (x+5)^2 - 25$ which fits the pattern $a(x-h)^2 + k$.

This is the form we were searching for.

Now, if we have a more general quadratic like $3x^2 - 9x + 5$, we first isolate the expression of the form $x^2 + bx$ upon which we will perform the completion. Observe:

Given $3x^2 - 9x + 5$, we rewrite as $3(x^2 - 3x) + 5$ by factoring the leading coefficient out of the first

two terms. Now, we take half the coefficient of x (inside the parentheses), square it and add it

(inside the parentheses) then subtract the same quantity symbolically (inside the parentheses), obtaining

$$3x^2 - 9x + 5 = 3(x^2 - 3x) + 5 = 3\left(x^2 - 3x + \left(-\frac{3}{2}\right)^2 - \frac{9}{4}\right) + 5.$$

Notice that the leading 3 is distributed inside the parentheses. To arrive at the desired form, we move the subtracted

quantity outside resulting in $3\left(x^2 - 9x + \left(-\frac{3}{2}\right)^2\right) - \frac{27}{4} + 5.$

We then identify the square and combine like terms to obtain

$$3\left(x - \frac{3}{2}\right)^2 - \frac{7}{4}.$$

Thus, $3x^2 - 9x + 5 = 3\left(x - \frac{3}{2}\right)^2 - \frac{7}{4}.$

This is a lengthy procedure that requires patience and practice, but it is necessary to understand and be able to perform it well in college mathematics. Indeed, it can be shown using this technique (with patience and care!) that the quadratic $ax^2 + bx + c$ can be written as

$$ax^2 + bx + c = a\left(x - \left(-\frac{b}{2a}\right)^2\right) + \frac{4ac - b^2}{4a}.$$

Thus the vertex of the generic quadratic function $f(x) = ax^2 + bx + c$ can be written in terms of a, b, and c as

$$\text{Vertex} = \left(-\frac{b}{2a}, \frac{4ac - b^2}{4a}\right).$$

In practice though, since the vertex is a point on the graph of $f(x) = ax^2 + bx + c$, it is decidedly more convenient to simply remember that the x-coordinate of the vertex is given by $x = -\frac{b}{2a}$ and compute the y-coordinate as $f(-\frac{b}{2a})$. In several days, we will use the quadratic formula: A formula that provides the solutions (even if they're not real) to any quadratic equation. It is derived from the above method of completing the square.

Day 13 Work Notes:

Do remember that if a quadratic function happens to come given in the form $f(x) = a(x-h)^2 + k$, we can immediately identify the vertex as the point (h, k). We never need to expand and use the vertex form shown yesterday. Indeed, that completely defeats the purpose of the work done on completing the square of the general quadratic $ax^2 + bx + c$.

Day 14 Work Notes:

Most often in doing mathematics, we are interested in working within the real number system. However, there are applications in the real world (pun intended!) that can make use of an expanded number system, namely the Complex Number System.

To begin, we recall that the square root of a negative number cannot be real: There are three types of real numbers, positive numbers, negative numbers, and zero. The square root of a number x, is that number, which when squared gives x. (As mentioned earlier, when x is positive, we require its principal square root to also be positive.) But what if x is negative? Then, there can be no real number that when squared gives x. For suppose $x < 0$ and $y^2 = x$. Then, if y were real, y would have to be positive, negative, or zero. But, squaring any of these types of numbers never results in a negative number. Hence, the square root of a negative number cannot be real.

So, mathematicians have extended the real number system as follows. We define $\sqrt{-1}$ to be a number we call i. Thus, i is some number so that $i^2 = -1$. Clearly, i is not real. As an example of how we handle other negative numbers, we write

$$\sqrt{-9} = \sqrt{9}\sqrt{-1} = \sqrt{9} \cdot i = 3i.$$

As you can check,

$$(3i)^2 = 3i \cdot 3i = 3 \cdot i \cdot 3 \cdot i = (3 \cdot 3) \cdot (i \cdot i) = 3^2 \cdot i^2 = 9 \cdot i^2 = 9 \cdot (-1) = -9.$$

Thus, $3i$ is the square root of -9 in this expanded system.

In general, the Complex Numbers, sometimes denoted by $\mathbb{C}$, are defined by

$$\mathbb{C} = \{a + bi : a, b \in \mathbb{R}\}.$$

For example, 3+2i is a complex number. It is important to recognize that the real numbers "live inside" the complex numbers. For, if we allow $b = 0$ and let a range over all the real numbers, we see that every real number a can be written in the form $a + 0i$ which is a special subset of $\mathbb{C}$. For information purposes only, the set of complex numbers of the form $0 + bi$ as b ranges over all the real numbers, are called the imaginary (or sometimes 'pure imaginary' numbers). If a and b are both non-zero, we call this number *non-real*. For example $3 + 2i$ is non-real.
We define operations in $\mathbb{C}$ as follows:

1. $(a + bi) \pm (c + di) = (a + c) + (b + d)i$

2. $(a + bi) \cdot (c + di) = (ac - bd) + (ad + bc)i$

3. $\dfrac{a + bi}{c + di} = \dfrac{ac + dc}{c^2 + d^2} + \dfrac{(cb - ad)}{c^2 + d^2} i$

Admittedly, these formulas are too cumbersome to memorize (except addition and subtraction which should appear rather natural). Instead, observe that multiplication can be thought of in the same way as multiplying two binomials (some people call this the FOIL method). Observe:

$$(a+bi) \cdot (c+di) = (ac) + (ad) \cdot i + (cb) \cdot i + (bd) \cdot i^2 \text{ (and since } i^2 = -1) = (ac - bd) + (ad + bc)i \text{ as above.}$$

For division, we can consider multiplying by 1 in the form of the conjugate of the demoninator divided by itself:

$$\frac{a+bi}{c+di} = \frac{a+bi}{c+di} \cdot \frac{c-di}{c-di} = \frac{(ac+dc) + (cb-ad)i}{c^2 + d^2}.$$

To put this in the desired form for complex numbers (i.e., $x + yi$), we simply split this into two pieces as

$$\frac{(ac+dc) + (cb-ad)}{c^2 + d^2} = \frac{ac+dc}{c^2+d^2} + \frac{(cb-ad)}{c^2+d^2}i \text{ as above.}$$

As mentioned previously, the method of completing the square can be used to derive the Quadratic Formula. This is a formula that gives solutions to the generic quadratic equation $ax^2 + bx + c = 0$ in terms of the coefficients $a, b,$ and c. It provides all solutions within the complex numbers. It is...

Quadratic Formula: The solutions to $ax^2 + bx + c = 0$ are given by $x = \dfrac{-b \pm \sqrt{b^2 - 4ac}}{2a}$.

In general, this formula should only be used as a last resort. If the equation comes given in the form $a(x - h)^2 + k = 0$, then the square-root method is ideal. It provides the non-real solutions readily as well (assuming there are non-real solutions). If the quadratic comes in the form $ax^2 + bx + c = 0$ and $ax^2 + bx + c$ can be factored easily, that is definitely the best strategy. If not in standard form and not easily factorable, then we apply the quadratic formula.

Day 15 Work Notes:

Remember: If a quadratic equation can be put in the form

$a(x - h)^2 + k = 0$, it is easily solved using the square-root method, even if the solutions are not real.

Linear equations are easily solved as we've seen. Recall that linear inequalities are inequalities that can be put in the form

$ax + b < 0$ or $ax + b \le 0$.

Solving these is much like solving a linear equation: Isolate x on one side of the inequality. One thing to remember, though, is that if you multiply (or divide) both sides of an inequality by a negative number, the inequality symbol changes direction.

Solving nonlinear inequalities is deceptively harder. While there are several variations on the below method, note that there are no real shortcuts for getting around this matter: In general, non-linear inequalities involve a much more elaborate process to obtain a solution.

To solve the non-linear inequality $f(x) < 0$ (or, $f(x) \le 0$ or, $f(x) > 0$ or, $f(x) \ge 0$) we construct a *sign-chart* for $f(x)$. To do this, we follow the guidelines below:

1. Move all terms to one side of the inequality and obtain 0 on the other side (if not already done).

2. Find the *partition numbers* for $f(x)$. That is, all values x such that $f(x) = 0$ or such that $f(x)$ is undefined (i.e. doesn't exist).

3. Plot these numbers on a number line indicating the reason you've plotted them: Write 0 above the values on the number line if it is because they make the function 0 and write u above them if they make the function undefined. This breaks up the number line into some number of (open) intervals.

4. Pick a test number in each interval and evaluate $f(x)$ at that test number.

5. Fill in the sign of the function (+ or –) above the number line corresponding to the appropriate interval depending on whether f evaluated at the test number is positive or negative.

6. Read the solution from the sign-chart: The function f is positive wherever the "+" signs are on the number line and $f(x) < 0$ on the intervals where the "–" signs are. Further $f(x) = 0$ wherever indicated above the number line.

7. Identify the solution intervals to the inequality in question.

As you can see, linear inequalities are deceptively simple to solve. The important thing to remember is that the above procedure provides the general method for solving any inequality which is not linear. (It even works for linear inequalities too; but, as mentioned earlier, linear inequalities are much-more straight forward to solve and there is no need for such elaborate procedures.)

Day 16 Work Notes:

Just as we perform arithmetic operations on numbers, we also do this for functions in mathematics. Given functions f and g both defined at some number x, we now define:

1.   Addition/subtraction of functions: $(f \pm g)(x) = f(x) \pm g(x)$

2.   Multiplication of functions: $(f \times g)(x) = f(x) \cdot g(x)$

3.   Division of functions: $\left(\dfrac{f}{g}\right)(x) = \dfrac{f(x)}{g(x)}$ provided $g(x) \neq 0$

These definitions are all rather natural it seems. However, the most important way of obtaining new functions from old is by the operation of composition of functions. This involves performing one function on an input, taking the resulting output, and applying the second function on that value. We denote this operation using the composition symbol "∘". That is we define the composition "f composed with g of x" to be

$(f \circ g)(x) = f(g(x)).$

As can easily be demonstrated, the order here is important. In general, $(f \circ g)$ is a different function than $(g \circ f)$. One idea that is used throughout the sciences and mathematics is that of taking something complicated and breaking it into simpler pieces. For us, we would like to be able to *decompose* a complicated function into simpler pieces. An example perhaps best illustrates the idea. Consider the function

$h(x) = \sqrt{x^2 + 1}.$

Let $g(x) = x^2 + 1$ and $f(x) = \sqrt{x}$. We claim $h = f \circ g$.

For, $(f \circ g)(x) = f(g(x)) = f(x^2 + 1) = \sqrt{x^2 + 1} = h(x).$

This notion of decomposing functions into simpler pieces is useful in mathematics. One thing to note is that the process is, in general, not unique. For example, if above we would have let $g(x) = x^2$ and $f(x) = \sqrt{x + 1}$, it would still be true that $(f \circ g)(x) = h(x)$ as you can check.

Day 17 Work Notes:

For some functions it is true that there exists a second function that "undoes" the first and vice-versa . These functions are said to be inverses of each other.

For example, the functions $f(x) = 5x + 8$ and $g(x) = \dfrac{x - 8}{5}$ are inverses of each other.

While there is quite a lot to say about inverses, we'll stick to some basic information for now. First, the mathematical definition of a function and its inverse:

Definition: The functions f and g are said to be inverses of each other provided:

1. $(f \circ g)(x) = x$ for every x in the domain of g

2. $(g \circ f)(x) = x$ for every x in the domain of f.

If g is the inverse of f, we use the notation f^{-1} to denote g. For example, from above, if

$$f(x) = 5x + 8, \; f^{-1}(x) = \frac{x - 8}{5}.$$

It must be noted that there are functions which do not have inverses. In fact, the only functions that have inverses are the *one-to-one functions*. While somewhat technical, a function is said to be one-to-one provided that if x_1, x_2 are distinct values in the domain of f (i.e., $f(x_1)$ and $f(x_2)$ exist and $x_1 \neq x_2$), then $f(x_1) \neq f(x_2)$. One way to visualize this is the *horizontal-line test*. It states that if there is any horizontal line which intersects the graph of a function in more than one point, then that function is not one-to-one. Equivalently, if every horizontal line intersects the graph of a function in at most one point, then the function is a one-to-one function.

Regarding finding the inverse of a function when it exists, the standard techique is as follows:

1.   Given a function $y = f(x)$, interchange the variables x and y.

2.   Solve the resulting equation for y in terms of x.

3.   If the resulting equation represents y as a function of x, then, that function is the inverse of f.

4.   If the resulting equation does not represent y as a function of x, then the original function does not

          have an inverse.

Some people find the above procedure confusing. Regarding Step 1, the only reason we perform this step is so that we end up with a "usual" looking function where x denotes the input and y denotes the output.

Other pertinent information regarding a function and its inverse includes the following: Suppose f and g are inverse functions. Then,

1.   The graphs of f and g are reflections of each other through the line $y = x$.

2.   If (x, y) lies on the graph of f, then (y, x) lies on the graph of g.

3.   The domain and range of f are precisely the range and domain, respectively, of g.

Day 18 Work Notes:

Recall that exponential functions are functions of the form $f(x) = a^x$ where a is a positive number other than 1. Thus, the functions $2^x, 3^x, e^x$, and $(\frac{1}{2})^x$ are all examples of exponential functions.

The number a is called the base of the exponential function. If $a > 1$, the exponential function is increasing. If $0 < a < 1$, the function is decreasing. One important aspect to note is that exponential functions only take on positive values. Indeed, the range of an exponential function is $(0, \infty)$. On the other hand, the domain of an exponential function is $\mathbb{R} = (-\infty, \infty)$.

Observe that the usual rules for exponents still apply even if the exponent is a variable. For example,

$$a^x \cdot a^x = a^{x+x} = a^{2x}.$$

and since $(a^x)^k = a^{kx}$, we have ,

$$a^x \cdot a^x = a^{x+x} = a^{2x} = (a^2)^x.$$

Many modeling problems including population growth or decline, radioactive decay, and compounding of interest all involve exponential functions. Frequently we encounter functions of the form

$$A(t) = A_0 a^{kt} \text{ for fixed constants } A_0, a, \text{ and } k.$$

In this form, A_0 represents an initial quantity and a^k represents the base of an exponential function. Thus if $a^k > 1$, the modeling function is increasing. Similarly, if $0 < a^k < 1$, then the modeling function is decreasing.

Finally it worth mentioning that the function $f(x) = e^x$ is called *the* exponential function (or, sometimes, the natural exponential function). Recall that e is an irrational number that arises naturally within mathematics. It is approximated by the decimal

$$e \approx 2.71828.$$

The natural exponential function has been described as one of the most important functions in all of mathematics.

Day 19 Work Notes:

For each exponential function $f(x) = a^x$, there exists a corresponding inverse function, called a logarithmic function of base a denoted by $g(x) = \log_a x$. To say this means more precisely that

1. $\log_a a^x = x$ for every x in the domain of a^x (i.e., all real numbers)

2. $a^{\log_a x} = x$ for every x in the domain of $\log_a x$ (i.e, all positive real numbers).

To make sense of the domain requirements above recall that the domain and range of a function and its inverse exactly *flip-flop*. Since exponential functions satisfy:

Domain: $(-\infty, \infty) = \mathbb{R}$, and Range: $(0, \infty)$,

the logarithmic functions satisfy

Domain: $(0, \infty)$, and Range: $(-\infty, \infty) = \mathbb{R}$.

There are two special cases of which we need to be aware. If we are using the logarithmic function with base 10, we write $\log x$ without indicating the base. This function is called the *common logarithmic function*. If we are using the logarithmic function with base e, we write $\ln x$. This is called the *natural logarithmic function*.

To gain understanding in the relationship between these two sets of functions, we frequently speak of *the logarithm of a number* or, simply, the *log* of a number. The logarithm of a number simply means the output value of a logarithmic function when evaluated at the number. People sometimes say that logarithms are exponents. Indeed, they are. For example $\log_2 x$ is the exponent we must raise 2 to to get x. For example, $\log_2 8 = 3$ since $2^3 = 8$. In general, a key relationship is as follows:

$$\log_b N = L \iff b^L = N.$$

The left-hand side above is said to be written in logarithmic notation. The right-hand side is said to be in exponential notation. Today, you will begin practice in converting between the two as a way to help understand the meaning of logarithms.

One final note to keep in mind is that exponential and logarithmic functions are one-to-one (they have to be as they are inverses of each other). In practical terms, this means that if $a^m = a^n$, then $m = n$. You will encounter this idea in the exercises. For example, to solve

$$5^{2x} = 125, \text{ we note that } 125 = 5^3. \text{ Thus, we have } 5^{2x} = 5^3 \implies 2x = 3 \implies x = \frac{3}{2}.$$

Thus, $x = \dfrac{3}{2}$ is the solution of the equation $5^{2x} = 125$.

Day 20 Work Notes:

Remember that a good way to solve equations which involve fractions is to multiply through the equation by the least common denominator of all denominators involved. You must verify your solutions though, as extraneous solutions may appear when multiplying by a variable quantity.

Polynomial functions are frequently encountered in mathematics. Indeed, the linear and quadratic functions we've seen are examples of polynomial functions. A polynomial function is a function that can be written in the form

$$a_n x^n + a_{n-1} x^{n-1} + a_{n-2} x^{n-2} + \ldots + a_1 x + a_0 \text{ where } n \text{ is a non-negative integer.}$$

The degree of the above polynomial is n (here, we assume $a_n \neq 0$). We call a_n the leading coefficient. For any $c \in \mathbb{R}$, except $c = 0$, the constant function $f(x) = c$ is a polynomial function of degree zero. While not universally agreed upon, many mathematicians include the constant function $f(x) = 0$ in polynomial functions but say that its degree is undefined.

Day 21 Work Notes:

It is noted that every nonconstant polynomial function "blows up" to $\pm\infty$ as x gets larger in a positive as well as in a negative sense. For example, for the function $f(x) = x^3$, as x gets larger in a positive sense, so does $f(x)$ as is illustrated by the graph below:

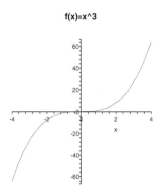

f(x)=x^3

Similarly, as x gets larger in a negative sense, so does $f(x)$. In this situation we write

$$x^3 \to \infty \text{ as } x \to \infty \text{ and}$$

$$x^3 \to -\infty \text{ as } x \to -\infty.$$

In general, given a nonconstant polynomial function, we sometimes desire to determine this "end-behavior". That is what happens as x gets larger in a positive sense (denoted by the symbols $x \to \infty$) and what happens as x gets larger in a negative sense (denoted by $x \to -\infty$).

One nice approach to detecting this end-behavior hinges on understanding the end-behavior of 4 nice functions. Namely, $\{y = x^2, y = -x^2, y = x^3, y = -x^3\}$. The graphs of these functions are given below.

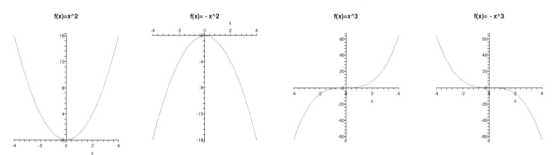

f(x)=x^2   f(x)= -x^2   f(x)=x^3   f(x)= -x^3

As you can see, every possible combination of end-behaviors is represented by one of these graphs. Given a nonconstant polynomial function we can determine its end-behavior by matching it with one of these four graphs. To do so, there are only two things we need to consider: the leading coefficient (whether it's positive or negative) and the degree (whether it's even or odd). There are 4 possible combinations of these features and each is represented by one of the four functions above. The promise is that the end-behavior of the function in question is identical to the end-behavior of the matching function above.

For example, suppose we are given the polynomial function

$$g(x) = -17x^5 + 2x^4 - 3x^3 + x$$

and wish to determine its end-behavior. Since the leading coefficient is negative and the degree is odd, we match it with $y = -x^3$ above since that is the only one of the four functions which has a negative leading coefficient and odd degree. Since

$$-x^3 \to -\infty \text{ as } x \to \infty, \text{ and}$$

$$-x^3 \to \infty \text{ as } x \to -\infty,$$

so does $g(x)$. That is the polynomial $g(x) = -17x^5 + 2x^4 - 3x^3 + x$ satisfies

$g(x) \to -\infty$ as $x \to \infty$, and

$g(x) \to \infty$ as $x \to -\infty$.

Polynomial division is another tool that remains essential for work in mathematics. Given a polynomial $p(x)$ and a non-zero divisor polynomial $d(x)$ of degree no more than that of $p(x)$, we wish to be able to find

$\dfrac{p(x)}{d(x)}$ or, equivalently in division notation, $d(x) \overline{)p(x)}$.

It is a mathematical fact that under these conditions, there exist unique polynomials $q(x)$ and $r(x)$ (the quotient and remainder polynomials, respectively) so that

$p(x) = q(x) \cdot d(x) + r(x)$ where the degree of $r(x)$ is less than the degree of $d(x)$ (or $r(x) = 0$).

The procedure for performing polynomial division is quite similar to long division with numbers, but slightly more abstract. One word of caution here: You may have encountered a shortcut called "synthetic division" for performing certain polynomial division problems. These division problems are quite special. Synthetic division can *only* be utilized when the divisor is of the form $x - c$ for some real number c. There are other elements to the method that are frequently forgotten or overlooked as well. So, while not completely irrelevant, unless you perform polynomial division frequently, our advice is to become proficient using long polynomial division and only spend effort on synthetic division once other pertinent skills are mastered.

Day 22 Work Notes:

Just as we classify integers as even or odd, we also have a classification for functions as even or odd. Note right away that most functions are neither though. Even and odd functions are characterized by certain types of symmetry in their graphs as well as by algebraic properties.

Even functions are defined by the characteristic that

$f(-x) = f(x)$ for every x in the domain of f.

Similarly, we say that a function $f(x)$ is *odd* provided that

$f(-x) = -f(x)$ for every x in the domain of f.

To test a function $y = f(x)$ for evenness or oddness, we compute $f(-x)$. If the result is $f(x)$, the function is even. If the result is $-f(x)$, the function is odd. If it is neither of these, then the function is neither even nor odd.

The names even and odd come from the symmetry exhibited by power functions x^n. If n is an even integer, the graph of $y = x^n$ is symmetric about the y-axis. Observe the graphs of $y = x^2$ and $y = x^4$ below.

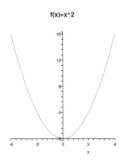

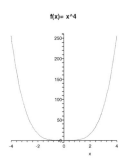

On the other hand, if the exponent n is an odd integer, the graph is said to be *symmetric with respect to the origin* or *symmetric about the origin*. This means that if the graph is reflected through one axis followed by a reflection through the other axis, then the graph maps precisely onto itself. Observe the graphs of $y = x^3$ and $y = x^5$ below:

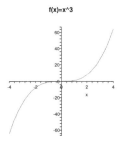

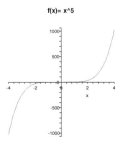

While the names even and odd originate from these notions, it should be pointed out that there are other functions which exhibit these types of symmetry as well. As examples, the function $y = \cos x$ is an even function and the function $y = \sin x$ is an odd function. We will not work with these functions in this book. Their graphs are given for observation purposes though. Observe that their graphs exhibit the types of symmetry described above.

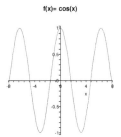

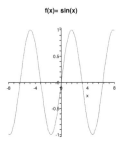

Just as we have done for polynomial functions, we sometimes desire to determine the "end-behavior" of a rational function. That is, given a rational function $q(x)$, what happens to $q(x)$ as x gets larger in a positive sense (denoted by the symbols $x \to \infty$) and what happens to $q(x)$ as x gets larger in a negative sense (denoted by $x \to -\infty$). (Recall that a rational function is a quotient of a polynomial function divided by another (non-zero) polynomial function.) For example,

$$q(x) = \frac{2x^3 - x^2 + 7}{-8x^3 + 4x} \text{ is a rational function.}$$

To determine the values that should be entered to make the statements

$$q(x) \to \underline{\quad} \text{ as } x \to \infty, \text{ and } q(x) \to \underline{\quad} \text{ as } x \to -\infty \text{ correct,}$$

we observe the degrees and leading coefficients of the numerator and denominator polynomials. There are three possible cases which determine our conclusions (one way to verify the following observations is to factor out the largest power of the variable in either the numerator or denominator. Here, we simply state the conclusions for reference). Given a rational function $q(x)$, the following hold:

1. If the degree of the numerator is less than the degree of the denominator, then always,

   $$q(x) \to 0 \text{ as } x \to \infty, \text{ and}$$

   $$q(x) \to 0 \text{ as } x \to -\infty$$

2. If the degree of the numerator is equal to the degree of the denominator, then consider the number c which is the ratio of the leading coefficient of the numerator divided by the leading coefficient of the denominator. Then, always,

   $$q(x) \to c \text{ as } x \to \infty, \text{ and}$$

   $$q(x) \to c \text{ as } x \to -\infty.$$

   Indeed, this turns out to be the "interesting case".

3. If the degree of the numerator is greater than the degree of the denominator, perform polynomial division to write the rational function as a polynomial + remainder term. The end-behavior of $q(x)$ is identical to the end behavior of the resulting polynomial term. (In fact, it is only the leading term of the quotient polynomial that is relevant, as that term determines the end-behavior of the quotient polynomial. With a little practice, you may be able to "see" this leading term without performing any polynomial division.)

Day 23 Work Notes:

It is frequently easier to manipulate mathematical objects involving radicals by first rewriting with rational exponents. For example to multiply

$$\sqrt{x} \cdot \left(\sqrt[3]{x} + \sqrt[4]{x} \right) \text{ we can rewrite as } x^{\frac{1}{2}} \left(x^{\frac{1}{3}} + x^{\frac{1}{4}} \right) = \left(x^{\left(\frac{1}{2} + \frac{1}{3} \right)} + x^{\left(\frac{1}{2} + \frac{1}{4} \right)} \right) = x^{\frac{5}{6}} + x^{\frac{3}{4}}.$$

If we wish to rewrite our answer back into radical notation, we have

$$\sqrt{x} \cdot \left(\sqrt[3]{x} + \sqrt[4]{x} \right) = \sqrt[6]{x^5} + \sqrt[4]{x^3}.$$

Day 24 Work Notes:

Remember that Pascal's triangle is a useful way for identifying the coefficients in multiplying a binomial by itself a number of times. For example, suppose we wish to expand

$$(a + b)^5.$$

We begin by constructing 6 lines of Pascal's triangle.

$$
\begin{array}{ccccccccccc}
 & & & & & 1 & & & & & \\
 & & & & 1 & & 1 & & & & \\
 & & & 1 & & 2 & & 1 & & & \\
 & & 1 & & 3 & & 3 & & 1 & & \\
 & 1 & & 4 & & 6 & & 4 & & 1 & \\
1 & & 5 & & 10 & & 10 & & 5 & & 1
\end{array}
$$

$$
\begin{array}{c}
\cdot \\
\cdot \\
\cdot
\end{array}
$$

The coefficients of the terms of the expansion are found in the sixth row of Pascal's triangle. The powers of the "a-terms" descend while the powers of the "b-terms" ascend. That is, we have,

$$(a + b)^5 = a^5 + 5a^4b + 10a^3b^2 + 10a^2b^3 + 5ab^4 + b^5.$$

While this is not the only way of determining the coefficients, it is quite useful for relatively low-degree factor expansion.

One final note, if it is a difference, such as $(a - b)^5$, the signs alternate beginning with a positive first term. That is,

$$(a - b)^5 = a^5 - 5a^4b + 10a^3b^2 - 10a^2b^3 + 5ab^4 - b^5.$$

Day 25 Work Notes:

There are many "formulas" encountered in mathematical applications. Having the skills necessary to rearrange formulas to solve for a given variable is valuable. Remember that a tool for working with exponents involves logarithms. The key property of logarithmic functions worth remembering is

$$\log_b N^c = c \cdot \log_b N.$$

The above property holds for any base logarithm. In the sciences, we almost exclusively use the natural logarithm, that is, the logarithmic function with base e. Recall that there is special notation for this base, namely ln. Thus, the above property for the natural logarithm function looks like

$$\ln N^c = c \cdot \ln N.$$

Also, entering complicated expressions in a calculator requires some care. We have to learn to think like a machine. The key area of concern is in order of operations. The calculator is programmed with these rules and will follow them. It is up to you, however, to ask the calculator to compute what you really want it to.

Day 26 Work Notes:

Factoring with negative exponents is a useful skill. It is frequently a much-more efficient way of simplifying expressions. For example, suppose we'd like to combine

$$\frac{a^3}{b^2} + \frac{b^5}{a^4}.$$

Now, while we certainly could get a common denominator and combine these two fractions, for more complicated expressions, that procedure can be cumbersome. Let's examine an alternate method here. We'll start by rewriting with negative exponents. We have,

$$a^3 b^{-2} + b^5 a^{-4}.$$

Now, before simplifying this expression, let's consider an example with positive exponents since the procedure works exactly the same way with negative exponents as it does with positive ones. Suppose we wish to simplify

$$a^3 b^6 + a^2 b^8.$$

We observe that both a and b occur on both sides of the "+" symbol. Thus, we can factor both from the expression. The only question is "To what power are we going to factor these terms?" As you likely recognize, we look for the smallest exponent on the a term and the smallest exponent on the b term and factor these out. Thus, we will factor out $a^2 b^6$. To determine what is left, we take what we had and subtract what we've taken out. That is,

$$a^3 b^6 + a^2 b^8 = a^2 b^6 \left(a^{(3-2)} b^{(6-6)} + a^{(2-2)} b^{(8-6)} \right)$$

and arrive at

$$a^3 b^6 + a^2 b^8 = a^2 b^6 \left(a + b^2 \right).$$

Now, the procedure with negative exponents works in exactly the same fashion. Again, we consider

$$a^3 b^{-2} + b^5 a^{-4}.$$

We choose the smallest exponent on the a factor (namely, -4) and the smallest exponent on the b factor (namely, -2) and we factor the term $a^{-4} b^{-2}$ out of the expression. To determine what is left over, we subtract just as we did with positive exponents: *what we had minus what we've taken out.* Thus, we arrive at

$$a^3 b^{-2} + b^5 a^{-4} = a^{-4} b^{-2} \left(a^{(3-(-4))} b^{(-2-(-2))} + a^{(-4-(-4))} b^{(5-(-2))} \right) = a^{-4} b^{-2} \left(a^7 + b^7 \right).$$

And so, we have, $a^3 b^{-2} + b^5 a^{-4} = a^{-4} b^{-2} \left(a^7 + b^7 \right) = \dfrac{a^7 + b^7}{a^4 b^2}.$

You will get a chance to practice this skill in today's exercise set.

Day 27 Work Notes:

As you've encountered earlier, in converting from exponential to logarithmic notation or vice-versa, the equivalence

$$\log_b N = L \iff b^L = N \text{ is quite useful.}$$

One additional aspect that this implies relates to the fact that for any admissible base b of a logarithmic function (the positive numbers other than 1), we have

$$b^0 = 1.$$

Applying the above equivalence, we have

$$\log_b 1 = 0.$$

Thus, for any choice base it is true that $\log_b 1 = 0$. We could also phrase this fact in either statement:

- The point $(0, 1)$ lies on the graph of every exponential function.
- The point $(1, 0)$ lies on the graph of every logarithmic function.

Reminder: If an inequality is not linear, a more elaborate process for solving the inequality may be required.

Day 28 Work Notes:

Often we are presented with the problem of finding solutions to multiple equations involving multiple variables. While a whole subject in and of itself, we will examine only the situation where we are presented with two linear equations involving two variables. To *solve this system* means to find values for the variables that satisfy both equations simultaneously.

Given two linear equations in two unknowns (i.e., two variables), there are three distinct possibilities regarding a solution to the system: There will either be zero, one, or infinitely many solutions. Given a linear equation in variables x and y, for example, $3x - 4y = 9$, we can regard the graph of this equation as a line in the x-y plane (provided the coefficients of x and y are not both zero). Given another equation (such as $x - 2y = 1$), we can like-wise consider the line determined by this equation. The solution to the system (if there is one) is represented by the intersection of these two lines. There are three possibilities for the intersection of two lines in a plane.

1. The lines do not intersect (i.e., they are parallel).

2. There is exactly one unique point of intersection.

3. The lines intersect in infinitely many points (i.e., the equations really represent the same line).

These three geometric scenariois correspond precisely to the possibilities for the solution of a linear system of two equations in two unknowns.

There are several methods that can be utilized to solve a system of two linear equations in two unknowns.

Method 1: Substitution-

If one of the equations can be used to solve for a variable in terms of the other easily (and resulting in nice numbers), this method is preferable. For example, consider the two equations posed above

$$3x - 4y = 9$$

$$x - 2y = 1.$$

Since we can write the second equation as

$$x = 2y + 1,$$

we can substitute the value for x in the first equation. That is

$$3(2y + 1) - 4y = 9.$$

This gives a linear equation involving only one variable. We then solve this equation

$$3(2y + 1) - 4y = 9 \implies 6y + 3 - 4y = 9 \implies 2y = 6 \implies y = 3.$$

To finish solving the system, once we know y we substitute into either equation and determine x. Clearly since we have x solved for in terms of y in the equation $x = 2y + 1$, this equation provides the easier calculation and provides

$$x = 2 \cdot 3 + 1 = 7.$$

Thus, the solution to the system appears to be

$$x = 7 \text{ and } y = 3.$$

We should always check that both equations are satisfied by this solution pair. Indeed, they are as you can check.

Method 2: Adding or subtracting equations to eliminate a variable-

In this situation, we may be able to easily eliminate one variable by generating a new equation by adding or subtracting the two equations. Consider

$$3x + y = 10$$

$$-5x - y = 12.$$

If we add the two equations we have

$$-2x = 22$$

so that $x = -11$. Once we know the value of one variable, we can substitute into either equation just like in Method 1 to obtain the value of the other. Thus, in this example, when $x = -11$, we have (using the first equation),

$$3(-11) + y = 10 \implies -33 + y = 10 \implies y = 43.$$

Again checking that this pair of values satisfies both original equations is good practice.

Method 3: Adding a multiple of one equation to another-

This approach is similar to Method 2. However, suppose adding or subtracting equations immediately does not eliminate a variable. Then, first we will multiply through the equations by values that provide the same coefficient of one variable yet opposite in sign. This will allow us to utilize the addition/subtraction technique of Method 2 to finish the solution process. Observe the equations

$$2x - 3y = -2$$

$$3x + 5y = 35.$$

To solve this system we have several choices. Here, we'll choose to try to eliminate the y variable as the signs are already opposite on the coefficients of the y-terms. To do so, we'll multiply through the first equation by 5 and through the second equation by 3. We obtain,

$$10x - 15y = -10$$

$$9x + 15y = 105.$$

Now we are in positition to utilize Method 2: Add the two equations to obtain

$$19x = 95 \text{ which implies } x = 5.$$

We then determine y by substituting $x = 5$ into either equation (the original or one of the multiple-generated equations). We'll use the first original equation $2x - 3y = -2$ to obtain

$$2(5) - 3y = -2 \Rightarrow 10 - 3y = -2 \Rightarrow -3y = -12 \text{ or } y = 4.$$

As you can check in the original system, this pair, $x = 5, y = 4$ is, indeed a solution to the given system.

Note that Methods 1 and 3 can always be used to determine the solutions to the type of systems we are considering. It frequently depends on the specific equations we encounter as to which method we utilize.

Finally, simply as examples of systems where there are no solutions or infinitely many solutions, we leave you with the two following systems:

The following system has no solution:

$$3x + y = 7$$
$$3x + y = 9.$$

It should be clear that there can be no numbers x and y so that $3x + y$ equals both 7 and 9 at the same time.

This system has infinitely many solutions:

$$-3x + y = 8$$
$$6x - y = -16.$$

Hint: To see this, write both equations as lines in the form $y = mx + b$.

Day 29 Work Notes:

There are several things to keep in mind for today's exercises. Recall that log functions are only defined on the set of positive real numbers. Also, recall that exponential functions (i.e., functions of the form $f(x) = a^x$ where $a > 0, a \neq 1$) only take on positive values.

And, given a rational function $q(x)$, in order to determine the values that should be entered to make the statements

$$q(x) \rightarrow \text{\_\_\_\_ as } x \rightarrow \infty, \text{ and } q(x) \rightarrow \text{\_\_\_\_ as } x \rightarrow -\infty$$

correct, we recall the three possible cases.

1. If the degree of the numerator is less than the degree of the denominator, then always,

   $$q(x) \rightarrow 0 \text{ as } x \rightarrow \infty, \text{ and}$$
   $$q(x) \rightarrow 0 \text{ as } x \rightarrow -\infty$$

2. If the degree of the numerator is equal to the degree of the denominator, then consider the number c which is the ratio of the leading coefficient of the numerator divided by the leading coefficient of the denominator. Then, always,

   $$q(x) \rightarrow c \text{ as } x \rightarrow \infty, \text{ and}$$
   $$q(x) \rightarrow c \text{ as } x \rightarrow -\infty.$$

3. If the degree of the numerator is greater than the degree of the denominator, determine the leading term of the quotient polynomial when performing polynomial division. The end-behavior of $q(x)$ is identical to the end behavior of the resulting quotient polynomial.

Day 30 Work Notes:

Congratulations. If you've worked through the previous 29 days of exercises diligently and have made an effort to fill in any gaps as needed, you should find today's exercise set relatively straightforward. Moreover, you should be confident that you have the skills necessary to begin studying college-level mathematics. Nice job.

PART III: Complete Answer Set

Solutions

Day 1

1 Daily Drill on Fundamentals: Multiplying Algebraic Expressions

1. $2x^2 + 3x$

2. $x^4 + 4x^2$

3. $x^2 + 3x - 10$

4. $x^3 - 2x^2 - 3x$

5. $6x^2 - 10x - 4$

6. $x^2 - 2x + 1$

2 Solving Algebraic Equations

1. $x = 5/2$

2. $x = 0$ or $x = 2/3$

3. $x = 0$ or $x = -7$ or $x = 3/4$

3 Factor Expansion

1. $x^2 - 9$

2. $4x^2 - 25$

3. $4x^2 - 12x + 9$

4. $x^3 + 12x^2 + 48x + 64$

5. $27x^3 + 54x^2 + 36x + 8$

6. $x^4 + 4x^3 + 6x^2 + 4x + 1$

4 Math To Go

1. True. If $a \cdot b = 0$, then $a = 0$ or $b = 0$.

2. False. Neither $x = 5$ nor $x = 2$ are solutions as they do not satisfy the equation. Indeed, rewriting the equation as $x^2 - 3x - 5 = 0$ and using the quadratic formula, it can be shown that the solutions are

$$x = \frac{3}{2} \pm \frac{\sqrt{29}}{2}$$

Solutions continued...

Day 2

1　Daily Drill on Fundamentals: Multiplying Algebraic Expressions

1.　$6x - 15x^2$

2.　$x^4 - 2x^3 + 2x^2 - 2x + 1$

3.　$x^2 + \dfrac{5}{2}x - \dfrac{3}{2}$

4.　$x^3 + 4x^2 - 9x - 36$

5.　$6x^3 - 7x^2 + 2x$

6.　$9x^2 - 6x + 1$

2　Solving Algebraic Equations

1.　$x = 5/3$

2.　$x = 3$ or $x = -5/3$

3.　$x = 2/3$ or $x = 0$ or $x = 4$

3　More Factor Expansion

1.　$x^3 - 6x^2 + 11x - 6$

2.　$x^3 + 5x^2 - 25x - 125$

3.　$27x^3 + 81x^2 + 81x + 27$

4.　$x^4 - 4x^3 + 6x^2 - 4x + 1$

5.　$x^5 + 5x^4 + 10x^3 + 10x^2 + 5x + 1$

6.　$1 - 4x + 6x^2 - 4x^3 + x^4$

4　Math To Go

Let x be the length and y denote the width of the rectangular corral.

1.　$y = 50 - x$

2.　$\{x : 0 < x < 50\}$ or, in interval notation, $(0, 50)$.

Solutions continued...

Day 3

1   Daily Drill on Fundamentals: Factoring

1. $(x)(x-1)$

2. $(x-3)(x+3)$

3. $(x^2)(x+2)$

4. $(x-2)(x+2)(x^2+4)$

5. $(x+1)(x^2-x+1)$

6. $(x+4)(x+5)$

2   Linear Equation Practice

1. slope $= 3$

2. $y = 3x$

3. $y = 0$ or, alternately, $(0,0)$.

3   Interval Notation

1. $(4, \infty)$

2. $(-\infty, 5) \cup (5, \infty)$

3. $(-\infty, 2] \cup (5, \infty)$

4. $(0, 10]$

5. $(-\infty, -4) \cup (4, \infty)$

6. $[-3, -1]$

4   Math To Go

1. 1850 sq. m

Solutions continued...

Day 4

1 Daily Drill on Fundamentals: Factoring

1. $(x^3)(x+1)(x^2-x+1)$

2. $(x-2)(x-3)$

3. $(x^2)(x^2+3)$

4. $(2x+1)(x-5)$

5. $(x-1)(x^2+x+1)$

6. $(x)(3x+1)(x-2)$

2 Linear Equation Practice

1. slope $= 2$

2. $y = 2x + 8$

3. $y = 8$ or, alternately, $(0, 8)$.

3 Interval Notation

1. $\{x : 5 \leq x \leq 10\}$

2. $\{x : -3 \leq x < 4\}$

3. $\{x : x < 0\}$

4. $\{x : x \leq 0\})$

5. $\{x : 1 < x < 2\}$

6. $\{x : 10 \leq x\}$

4 Math To Go

1. 45 m x 90 m

Solutions continued...

Day 5

1   Daily Drill on Fundamentals: Solving Linear Equations

1. $x = 4$ 4. $a = -2$

2. $x = 8$ 5. $t = 1$

3. $x = 12$ 6. $x = 2, y \in \mathbb{R}$

2   Linear Equations

1. slope $= 5$

2. slope $= 12$

3. slope $= \frac{4}{3}$

3   Interval Notation

1. $(-\infty, -5]$ 4. $[-4, 0)$

2. $(-\infty, 0) \cup (0, \infty)$ 5. $(-5, 5)$

3. $(-\infty, 0] \cup [2, \infty)$ 6. $(-\infty, -15] \cup [-5, \infty)$

4   Math To Go

1.

$$\text{radius } = \frac{68}{2\pi} \text{ cm } = \frac{34}{\pi} \text{ cm} \approx 10.8 \text{ cm.}$$

2.

$$\text{radius } = r = \frac{C}{2\pi}.$$

Solutions continued...

Day 6

1   Daily Drill on Fundamentals: Exponents and Notation

1. 9

2. 9

3. 4

4. undefined in $\mathbb{R}$

5. 9

6. -27

2   Rational Exponents

1. $16^{3/2}$

2. $9^{3/2}$

3. $a^{m/n}$

3   More Rational Exponents

1. True

2. False

3. True

4. True

5. True

6. False

4   Math To Go

1.

$$\text{radius} = \sqrt[3]{\frac{3 \cdot 33510}{4\pi}} \text{ cm} \approx 20 \text{ cm.}$$

2.

$$\text{radius} = r = \sqrt[3]{\frac{3V}{4\pi}}$$

Solutions continued...

Day 7

1 Daily Drill on Fundamentals: Solving Linear and Linear-like Equations

1. $x = \dfrac{1}{5}$

2. $x = \dfrac{13}{3}$

3. $x = \dfrac{2}{3}$

4. $x = -\dfrac{2}{3}$

5. $x = -2$

6. No solutions

2 Function Notation Practice

1. 6

2. 4

3. $\dfrac{12}{\sqrt{t} - 2}$

3 Domain of a Function

1. $(-\infty, 3) \cup (3, \infty)$

2. $[3, \infty)$

3. $(-\infty, -2) \cup (-2, 3) \cup (3, \infty)$

4. $[2, 5) \cup (5, \infty)$

5. $(-\infty, \infty) = \mathbb{R}$

6. $(-\infty, \infty) = \mathbb{R}$

4 Math To Go

1. 21.8 m

Solutions continued...

Day 8

1 Daily Drill on Fundamentals: Equivalent Expressions

1. No. Their domains are different.

2. No. Their domains are different.

3. Yes. They both provide the same value for all real numbers.

2 Function Notation Practice

1. 3

2. $t^2 - t + 3$

3. $(x + h)^2 - (x + h) + 3 = x^2 + 2xh + h^2 - x - h + 3$

3 Domain of a Function

1. $(-\infty, -4) \cup (-4, 4) \cup (4, \infty)$

2. $[1/2, \infty)$

3. $(-\infty, -1) \cup (-1, 1) \cup (1, \infty)$

4. $(-\infty, 5/2) \cup (5/2, \infty)$

5. $(-\infty, \infty) = \mathbb{R}$

6. $(-\infty, 2)$

4 Math To Go

1. Five years after being transplanted, the sapling is ten feet tall.

2. Ten minutes after injection, there is 1/2 mg/dL of drug in the patient's blood.

Solutions continued...

Day 9

1 Daily Drill on Fundamentals: Factoring

1. $(3x-1)(x+6)$

2. $-(x-6)(x+5)$

3. $(x+3)(x-3)(x^2+4)$

4. $12(x-2)(x^2+2x+4)$

5. $(x+1)(x-\sqrt{3})(x+\sqrt{3})$

6. $(x)(x-\sqrt{3})(x+\sqrt{3})(x^2+2)$

2 Linear Equation Practice

1. $y = 3x + 1$

2. Graph the line with slope 3 which passes through $(1,4)$ in the $x-y$ plane.

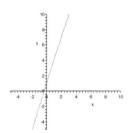

3 Domains of functions

1. $\{r : r > 0\}$ or $(0, \infty)$

2. $[0, \sqrt{6}]$

3. $(0, 8)$

4 Math To Go

1. $t = 5/8$ seconds

Solutions continued...

Day 10

1   Daily Drill on Fundamentals: Factoring

1. $(3x + 2)(2x - 1)$ 4. $(x - 3)(x^2 + 3x + 9)$

2. $(x - \sqrt{5})(x + \sqrt{5})$ 5. $(x + 2)(x - \sqrt{3})(x + \sqrt{3})$

3. $(3x - 2)(x + 3)$ 6. $(12x - 1)(x + 4)$

2   Quadratic Equation Practice

1. $x = \pm\sqrt{2}$

2. $x = \pm\frac{5}{3}$

3. $x = 12$ or $x = -6$

3   The Graph of Quadratic Equation: The Parobala

1. $x = \pm 3$ or, alternately   4.   No x-intercepts.

        $(-3, 0)$ and $(3, 0)$

2. $x = 5/3$ and $x = -1$ or, alternately   5. $x = 5$ or, alternately $(5, 0)$

        $(5/3, 0)$ and $(-1, 0)$

3. $x = 3 \pm \sqrt{2}$ or, alternately   6. $x = -3$ and $x = 7$ or, alternately

        $(3 - \sqrt{2}, 0)$ and $(3 + \sqrt{2}, 0)$ $(-3, 0)$ and $(7, 0)$

4   Math To Go

1.   1.25 seconds

Solutions continued...

Day 11

1   Daily Drill on Fundamentals: Factoring

1. $3(2x - 1)(x + 2)$

2. $(x^2)(x - 3)(x + 2)$

3. $(x + 1)(x^2 - x + 1)$

4. $(x + \sqrt[3]{2})(x^2 - \sqrt[3]{2}x + \sqrt[3]{4})$

5. $(2x^2 + 1)(x + 1)(x - 1)$

6. $(x^5)(x - \sqrt{3})(x + \sqrt{3})$

2   Drill on Domains

1. $(-\infty, -4) \cup (-4, 5) \cup (5, \infty)$

2. $[4/3, 9/2) \cup (9/2, \infty)$

3. $(-\infty, \infty) = \mathbb{R}$

3   The graph of a quadratic function

1. 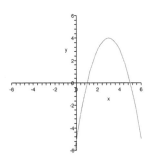   2.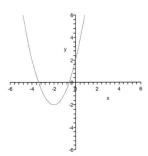

4   Math To Go

1. The maximum height is 22 ft and the lime reaches this height at time $t = 2$ seconds.

Solutions continued...

Day 12

1   Daily Drill on Fundamentals: Completing the Square

1. $(x + 3)^2 + 3$

2. $(x - 4)^2 - 26$

3. $2(x - 1)^2 - 14$

2   The Vertex of a Parabola

1. $(5, 7)$

2. $(1, 5)$

3. $(3/2, 11/4)$

3   Graphs of Quadratic Functions

1. Maximum

2. Maximum

3. Minimum

4   Math To Go

1. The lemon raches a maximum height of 31 feet at time $t = 1.25$ seconds.

Solutions continued...

Day 13

1 Daily Drill on Fundamentals: Completing the Square

1. $(x + \frac{1}{2})^2 + \frac{3}{4}$

2. $2(x - \frac{5}{4})^2 + \frac{31}{8}$

3. $-5(x - \frac{7}{10})^2 + \frac{309}{20}$

2 The Vertex of a Parabola

1. $\left(\dfrac{1}{2}, -\dfrac{5}{4} \right)$

2. $\left(-\dfrac{5}{2}, \dfrac{37}{4} \right)$

3. $\left(\dfrac{4}{3}, \dfrac{44}{3} \right)$

3 Maximum and Minimum Values of Quadratic Functions

1. The minimum is $\frac{17}{4}$.

2. The maximum is $\frac{55}{2}$.

3. The maximum is 0.

4 Math To Go

1. 58 miles per hour, 42 mpg

Solutions continued...

Day 14

1   Daily Drill on Fundamentals: Imaginary Numbers

1. $6i$ or $0 + 6i$

2. $\sqrt{3}i$ or $0 + \sqrt{3}i$

3. $3 - \sqrt{5}i$

4. $10 - 11i$

5. $-i$ or $0 - i$

6. $\dfrac{3}{10} + \dfrac{1}{10}i$

2   Quadratic Equation Practice

1. $-\dfrac{1}{2} \pm \dfrac{\sqrt{3}}{2}i$

2. $\dfrac{1}{3} \pm \dfrac{\sqrt{2}}{3}i$

3. $-1, 4$

3   The Discriminant: $b^2 - 4ac$

1. Two real solutions.

2. One real solution.

3. Two non-real solutions.

4   Math To Go

1. Downward. If it opened upward, there would be two x-intercepts giving two real solutions.

Solutions continued...

Day 15

1   Daily Drill on Fundamentals: Domains of Functions

1. $(-\infty, \infty) = \mathbb{R}$

2. $(-\infty, 4) \cup (4, \infty)$

3. $(-\infty, -4) \cup (-4, 5) \cup (5, \infty)$

4. $(-\infty, -1/2) \cup (-1/2, 2) \cup (2, \infty)$

5. $(-\infty, \infty) = \mathbb{R}$

6. $(-\infty, -2) \cup (-2, 2) \cup (2, \infty)$

2   Quadratic Equation Practice

1. $x = 4 \pm \sqrt{\dfrac{5}{3}}$

2. $x = -2 \pm \dfrac{3}{2}$

3. $x = \dfrac{7}{3} \pm \dfrac{2}{3} = \{3, 5/3\}$

3   A Parabola: The Graph of a Quadratic Function

1. $(4, 8)$

2. $(3, -9)$

3. $\left(\dfrac{9}{5}, \dfrac{76}{5}\right)$

4   Math To Go

1. Estimate: $(-\infty, 1) \cup (5, \infty)$. Solve: $(-\infty, 1) \cup (5, \infty)$.

Solutions continued...

Day 16

1   Daily Drill on Fundamentals: Combining Functions

1. $\sqrt{2x+3} + \dfrac{x}{x-1}$

2. $\dfrac{x\sqrt{2x+3}}{x-1}$

3. $\dfrac{\sqrt{2x+3}}{\frac{x}{x-1}} = \dfrac{(x-1)\sqrt{2x+3}}{x}$

    provided $x \neq 1$

4. $\sqrt{\dfrac{2x}{x-1} + 3}$

5. $\dfrac{\sqrt{2x+3}}{\sqrt{2x+3}-1}$

6. x provided $x \neq 1$

2   More composition

1. $\dfrac{1}{(5x^3-6)^2+3} = \dfrac{1}{25x^6 - 60x^3 + 39}$

2. $5\left(\dfrac{1}{x^2+3}\right)^3 - 6 = \dfrac{-6x^6 - 54x^4 - 162x^2 - 157}{x^6 + 9x^4 + 27x^2 + 27}$

3   Decomposing Functions

Answers vary. One set of possible answers:

1. $f(x) = 5x - 9, g(x) = x^3$

2. $f(x) = \sqrt{x}, g(x) = x - 7$

3. $f(x) = \dfrac{3}{x}, g(x) = x^4 + 2$

4. $f(x) = \dfrac{1}{\sqrt{x}}, g(x) = x - 5$

4   Math To Go

1. Compute $(1.0825)(129 - (.1)(129)) \approx \$125.68$

Solutions continued...

Day 17

1   Daily Drill on Fundamentals: Composition of Inverses

1. $(f \circ g)(x) = x, (g \circ f)(x) = x$ 3. $(f \circ g)(x) = x, (g \circ f)(x) = x$

2. $(f \circ g)(x) = x, (g \circ f)(x) = x$ 4. $(f \circ g)(x) = x, (g \circ f)(x) = x$

2   Inverses

1. $f^{-1}(x) = \dfrac{x + 8}{5}$ 2. $f^{-1}(x) = 6 - 2x^3$

3   Inverse Meaning and Notation

1. 3   3. 3
2. 0   4. 0

4   Math To Go

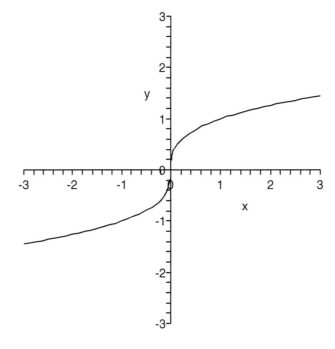

1.

Solutions continued...

Day 18

1   Daily Drill on Fundamentals: Factoring (Exponential Function Emphasis)

1. $2^x(1+x)$

2. $e^x(x+1+x^2)$

3. $e^x(e^x-1)$

4. $(e^x-3)(e^x+2)$

5. $x \cdot 3^x \cdot (x-1)^2$

6. $5^x(-8x)$

2   Solving Equations

1. $x=5$

2. $x=4$

3. $x=5$ or $x=-4$

3   More Factor Expansion

1. $5x^2-17x+6$

2. $2^{2x}-3\cdot 2^x-4$ or $4^x-3\cdot 2^x-4$

3. x^2+6x+9

4. $64-48x+12x^2-x^3$

5. $e^{3x}+3e^{2x}+3e^x+1$

6. $x^4-4x^3+6x^2-4x+1$

4   Math To Go

1. Increasing. The base of the exponential function $(1.012)^t$ is larger than 1.

Solutions continued...

Day 19

1   Daily Drill on Fundamentals: Exponential vs. Logarithmic Notation

1. $\log_3 81 = 4$

2. $\log_2 32 = 5$

3. $\log_{\frac{1}{2}} \dfrac{1}{8} = 3$

4. $\log_3 34 = L$

5. $\log_8 102 = N + 4$

6. $\log_b N = L$

2   Solving Equations

1. $x = 2$

2. $x = 5$

3. $x = -2$

3   More Form Conversion

1. $\log_5 100 = 2x$

2. $\log_{x+3} 19 = 4$

3. $\log_{\frac{1}{3}} 12 = x + 1$

4. $\log_{4-x} 1000 = 3$

5. $\log_{5x+1}(y + 6) = \dfrac{1}{2}$

6. $\log_{\frac{2}{3}}(y + 3) = x + 1$

4   Math To Go

1. $t = 5$ seconds

2. $t = 8$ minutes

Solutions continued...

Day 20

1 Daily Drill on Fundamentals: Multiplying Algebraic Expressions

1. $2x^3 - 18x$

2. $x^5 - x^4 + x^3$

3. $3x^4 + x^3 + 15x^2 + 5x$

4. $x^5 - 30x^4 + x^3 - 30x^2$

5. $15x^2 - 22x + 8$

6. $x^5 - 5x^4 + 10x^3 - 10x^2 + 5x - 1$

2 Solving Algebraic Equations

1. $x = \dfrac{1 \pm \sqrt{5}}{2}$

2. $x = 0$ or $x = \dfrac{2}{3}$

3. $x = \dfrac{2 \pm \sqrt{19}}{3}$

3 The Degree and Leading Coefficient of a Polynomial

1. Degree: 3; Leading Coefficient: 3
2. Degree: 2; Leading Coefficient: -5
3. Degree: 4; Leading Coefficient: 6

4. Degree: 3; Leading Coefficient: $-\frac{1}{2}$
5. Degree: 99; Leading Coefficient: 17
6. Degree: 0; Leading Coefficient: 3

4 Math To Go

1. Degree: 14, Leading Coefficient -243

2. Degree: 21, Leading Coefficient 24

Solutions continued...

Day 21

1 Daily Drill on Fundamentals: Polynomial and Rational Functions

1. $(-\infty, \infty) = \mathbb{R}$

2. $\left(-\infty, \dfrac{7}{2}\right) \cup \left(\dfrac{7}{2}, \infty\right)$

3. $(-\infty, \infty) = \mathbb{R}$

4. $(-\infty, -4) \cup (-4, 3) \cup (3, \infty)$

5. $(-\infty, \infty) = \mathbb{R}$

6. $(-\infty, \infty) = \mathbb{R}$

2 Solving Algebraic Equations

1. $x = -\dfrac{5}{2}$

2. $x = -\dfrac{2}{3}$ or $x = \dfrac{1}{2}$

3. $x = -1$ or $x = 6$

3 Polynomial Features: End-Behavior

1. $f(x) \to \infty$ as $x \to \infty$

   $f(x) \to -\infty$ as $x \to -\infty$

2. $f(x) \to -\infty$ as $x \to \infty$

   $f(x) \to -\infty$ as $x \to -\infty$

3. $f(x) \to \infty$ as $x \to \infty$

   $f(x) \to \infty$ as $x \to -\infty$

4. $f(x) \to -\infty$ as $x \to \infty$

   $f(x) \to \infty$ as $x \to -\infty$

4 Math To Go

1. $x^2 + x - 2$

Solutions continued...

Day 22

1   Daily Drill on Fundamentals: Even and Odd Functions

1. Even

2. Odd

3. Neither

4. Even

5. Neither

6. Odd

2   Solving Algebraic Equations

1. $x = -\dfrac{3}{2}$ or $x = 4$

2. No solutions

3. $x = \pm\dfrac{1}{6}$

3   Rational Functions: End-Behavior Again

1. $f(x) \to 0$ as $x \to \infty$

   $f(x) \to 0$ as $x \to -\infty$

2. $f(x) \to -3$ as $x \to \infty$

   $f(x) \to -3$ as $x \to -\infty$

3. $f(x) \to 1$ as $x \to \infty$

   $f(x) \to 1$ as $x \to -\infty$

4. $f(x) \to \infty$ as $x \to \infty$

   $f(x) \to \infty$ as $x \to -\infty$

4   Math To Go

1. $x^2 - x$

Solutions continued...

Day 23

1 Daily Drill on Fundamentals: Multiplying Algebraic Expressions

1. $3x^3 + 2x^2 - x$

2. $4x^2 - 25$

3. $x^3 - 5x^2 + 2x + 8$

4. $4x^{\frac{7}{2}} + x^{\frac{5}{2}} - 12x^{\frac{3}{2}} - 3x^{\frac{1}{2}}$

5. $2x^{\frac{3}{2}} + 2x + 7x^{\frac{1}{2}} + 7$

6. $x + 2$

2 Solving Algebraic Equations

1. $x = \dfrac{37}{11}$

2. $x = -4$ or $x = 5$

3. $x = -2$

3 More Factor Expansion

1. $2x^5 + 7x^4 - 4x^3$

2. $-2x + 10x^{\frac{1}{2}}y - 12y^2$

3. $x^5 - x^4 + 4x^3 - 4x^2$

4. $x^3 - 3x^2y + 3xy^2 - y^3$

5. $27x^3 - 27x^2y + 9xy^2 - y^3$

6. $x^4 + 4x^3y + 6x^2y^2 + 4xy^3 + y^4$

4 Math To Go

1. True. If $a \cdot b = 0$, then $a = 0$ or $b = 0$.

2. True. If $a \cdot b = 0$, then $a = 0$ or $b = 0$. If $x^2 + y^2 = 0$, then $x = 0 = y$. If $x^2 - y^2 = 0$, then $x = \pm y$.

Solutions continued...

Day 24

1 Daily Drill on Fundamentals: Multiplying Algebraic Expressions

1. $12x^{\frac{3}{2}} - 6x^{\frac{1}{2}}$

2. $x^4 - 2x^3y + x^2y^2 + x^2y - 2xy^2 + y^3$

3. $\dfrac{2x^2 + xy - y^2}{2}$ or

$x^2 + \dfrac{1}{2}xy - \dfrac{1}{2}y^2$

4. $x^3 + x^2z - xy^2 - y^2z$

5. $6x^2z - 5xyz + y^2z$

6. $9x^2 - 6xy + y^2$

2 Solving Algebraic Equations

1. $x = \pm 3$

2. $x = 9$

3. $x = \pm 4$ or $x = -16$ or $x = \dfrac{3}{4}$

3 More Factor Expansion

1. $x - 6x^{\frac{2}{3}} + 11x^{\frac{1}{3}} - 6$

2. $x^3 + x^2y - xy^2 - y^3$

3. $u^3 - 6u^2v + 12uv^2 - 8v^3$

4. $u^4 + 4u^3v + 6u^2v^2 + 4uv^3 + v^4$

5. $x^5 + 5x^4y + 10x^3y^2 + 10x^2y^3 + 5xy^4 + y^5$

6. $y^4 - 4y^3x + 6y^2x^2 - 4yx^3 + x^4$

4 Math To Go

1. $xy = 12600$

2. $2x + 2y = 500$

3. 180 ft x 70 ft

Solutions continued...

Day 25

1   Daily Drill on Fundamentals: Algebraic Expressions

1. $h = \dfrac{3V}{\pi r^2}$

2. $l = \dfrac{P - 2w}{2}$ or $l = \dfrac{P}{2} - w$

3. $h = \sqrt{\left(\dfrac{A}{\pi r}\right)^2 - r^2}$

4. $i = \sqrt[n]{\dfrac{A}{P}} - 1$

5. $n = \dfrac{\ln\left(\frac{A}{p}\right)}{\ln(1 + i)}$

6. $P = \dfrac{iV}{(1 + i)^n - 1}$

2   Evaluating Expressions using a Calculator

1. 648.47

2. 397077.36

3. 421.65

3   Basic Formulas

1. 54 sq. cm

2. 28 cm

3. 5π sq. cm

4. 45.5 sq. cm

5. 30 sq. cm

6. Radius = r = 4 cm

4   Math To Go

1. 96π cu. ft

Solutions continued...

Day 26

1   Daily Drill on Fundamentals: Factoring (Negative Exponent Emphasis)

1. $\dfrac{a^4 + b^7}{ab^5}$

2. $\dfrac{xy + 1}{x^3 y^4}$

3. $\dfrac{(1 - x^4)\sqrt{2x - 1}}{x^3}$

4. $\dfrac{y^6 - x^4}{x^3 y}$

5. $\dfrac{a^2 b^7 - a^7 + b^2}{a^5 b^4}$

6. $\dfrac{(x^2 - x - 1)^6 (5x + 3)^4 + 1}{(x^2 - x - 1)^3 (5x + 3)^6}$

2   Linear Equation Practice

1. Slope $= m = -6$

2. $y + 1 = -4(x - 3)$ or $y = -4x + 11$

3. $y = 6$ or $(0, 6)$

3   Combining and Simplifying Expressions

1. $\dfrac{1 + 2x}{x^2}$

2. $\dfrac{4x^2 - x}{x^2 - 2x - 8}$

3. $\dfrac{-2x^2 - 7x - 4}{x^2 - x - 20}$

4. $\dfrac{\sqrt{x - 7} + x}{x - 7}$

5. $\dfrac{3y^4 - x}{xy^5}$

6. $\dfrac{x^4 + y^6}{xy^4}$

4   Math To Go

1.

$$3x^2(2x^2 - 9x)^{-1} - (4x^4 - 9x^3)(2x^2 - 9x)^{-2} = (2x^2 - 9x)^{-2}[3x^2(2x^2 - 9x) - (4x^4 - 9x^3)] =$$

$$\frac{6x^4 - 27x^3 - 4x^4 + 9x^3}{(2x^2 - 9x)^2} = \frac{2x^4 - 18x^3}{(2x^2 - 9x)^2}$$

Solutions continued...

Day 27

1 Daily Drill on Fundamentals: Exponential vs. Logarithmic Notation

1. $2^4 = 16$

2. $5^{-3} = \dfrac{1}{125}$

3. $10^3 = 1000$

4. $e^0 = 1$

5. $b^L = N$

6. $\left(\dfrac{1}{3}\right)^{-2} = 9$

2 Solving Inequalities

1. $(-\infty, 6]$

2. $(-\infty, -1) \cup (7, \infty)$

3. $(-\infty, -5] \cup (1, 6]$

3 More Form Conversion

1. 3

2. -1

3. 2

4. 3

5. -3

6. 0

4 Math To Go

1. ≈ 63 times stronger.

Solutions continued...

Day 28

1   Daily Drill on Fundamentals: Multiplying Algebraic Expressions

1. $x^{\frac{7}{2}} - x^{\frac{1}{2}}$

2. $(3x-4)^{\frac{5}{6}}$ or $\sqrt[6]{(3x-4)^5}$

3. $x^{-\frac{1}{6}} - 3x^{-\frac{2}{3}} + x^{\frac{1}{2}} - 3$

4. $x^3 y^{\frac{5}{3}} - x^4 y$

5. $x + 2x^{\frac{1}{2}} + 1$

6. $x - 3x^{\frac{2}{3}} + 3x^{\frac{1}{3}} - 1$

2   Solving Algebraic Equations

1. $x = \dfrac{8}{3}$

2. $x = 3$

3. $x = 0$ or $x = 7$

Note: $x = 1$ is not a solution as the radical is not defined in $\mathbb{R}$ when $x = 1$.

3   Systems of Equations

1. $x = 5, y = 8$

2. $x = 3, y = -1$

3. $x = -2, y = 1$

4   Math To Go

1. $x = -40$ degrees

Solutions continued...

Day 29

1   Daily Drill on Fundamentals: Domains

1. $(-\infty, \infty) = \mathbb{R}$

2. $[1, 10) \cup (10, \infty)$

3. $(0, \infty)$

4. $(4, \infty)$

5. $(-\infty, \infty) = \mathbb{R}$

6. $(-\infty, 0) \cup (0, \infty)$

2   Solving Inequalities

1. $[23/2, \infty)$

2. $[0, 1]$

3. $(-\infty, \infty) = \mathbb{R}$

3   More End-Behavior

1. $f(x) \to -\infty$ as $x \to \infty$

   $f(x) \to \infty$ as $x \to -\infty$

2. $f(x) \to \infty$ as $x \to \infty$

   $f(x) \to 4$ as $x \to -\infty$

3. $f(x) \to 0$ as $x \to \infty$

   $f(x) \to 0$ as $x \to -\infty$

4. $f(x) \to -\infty$ as $x \to \infty$

   $f(x) \to -\infty$ as $x \to -\infty$

4   Math To Go

1. $x^2 - 2x + 1$

Solutions continued...

Day 30

1 Daily Drill on Fundamentals: Interval Notation

1. $(3, 10]$

2. $(-2, \infty)$

3. $(-\infty, 0] \cup (6, \infty)$

4. $[-2, 3)$

5. $(-\infty, -3) \cup (-3, 3) \cup (3, \infty)$

6. $(-\infty, -7) \cup (3, \infty)$

2 Equations of Lines

1. $y = -3x + 9$

2. $y = 5x - 7$

3. $y = -2x - 7$

3 Factoring

1. $(x + 3)(x + 4)$

2. $x^3(x - 9)(x + 9)$

3. This expression is irreducible over $\mathbb{R}$.

4. $(x - 1)^3$

5. $(x - 1)(x^2 + x + 1)$

6. $xe^x(1 - 3e^x)$

4 Math To Go

1. Kim must score 92 or better on the final to earn an A.

About the author:

M.J. Sanders is a professor of mathematics. He received a Ph.D. in Mathematics from the University of Tennessee for his work in geometric topology. He has taught thousands of college students both introductory and advanced mathematics courses over the last 25 years. He lives in Savannah, Georgia.

CPSIA information can be obtained
at www.ICGtesting.com
Printed in the USA
LVHW061341110420
653076LV00037B/1194